FINAL GIRL REDUX:

A RELAXED EXPLORATION OF WOMEN'S PLACE IN HORROR MOVIES

Final Girl Redux: A Relaxed Exploration of Women's Place in
Horror Movies
Copyright © Friendly Shark Publishing

Cover illustration designs by: Kimberly Pinzon
Contact: Kimberly.pinzon@gmail.com

First paperback edition
ISBN: 979-8-9895941-0-8

Contents

Dedicated to the girls and women who love horror, who peek through their fingers at the scary scenes, who scream at the characters "Don't go in there!", who laugh their way through unrealistic gore, and everyone in between.

Seize the genre and never let it go.

By its very nature, this book is full of

SPOILERS.

Proceed at your own risk as I will not be responsible for you reading about what happens in a movie that you have not seen.

For a comprehensive list of every movie mentioned or discussed in the following pages, find the "Movies Referenced" section at the end of this book.

Introduction

"It is women who love horror. Gloat over it. Feed on it. Are nourished by it. Shudder and cling and cry out – and come back for more."

Bela Lugosi

Jaws (1975) changed my life in three significant ways, and not just because I was shown the entire series in one night by a (maybe) well-intentioned aunt at the age of eight. My sister, who participated in the movie trauma, was only six so I will leave the judgment up to the reader.

The first way my life was changed was, predictably, I developed an intense fear of open bodies of water and the animals in them. Not just the ocean, no, my fear generalized to anything that might contain aquatic life (including swimming pools) and all of that aquatic life. Except, oddly, for sharks, leading into the second way *Jaws* affected me: I found an immense love and fascination for the creatures which were ultimately vilified and reviled due to the *Jaws* movies. Peter Benchley, the author of the book which spawned the movie, would later devote much of his life to the conservation of sharks and attempting to repair the public image of the animals.

Final Girl Redux

The third, most important way that *Jaws* affected my life, and the most relevant to this work, is that it unknowingly planted the seeds of my future love of horror.

I don't remember much of my actual memories of watching the movies, but I remember a persistent fear that lingered and caused nightmares for weeks even though I didn't live near the ocean and my risk of shark attack was essentially zero. But there were moments when I would remember a specific scene of the movie, or the general terror of Quint, Brody, and Hooper as the shark tears apart their boat. I didn't realize at the time that that terror would become somewhat of an addiction, and lead to a constant search for that next terror high when I became old enough to choose my own movies and escaped the clutches of parents who refused to let me watch anything that might contain the slightest bit of nudity, which included much of the sex saturated horror genre.

With *Jaws* safely in the past (or so I thought), my next brush with horror happened when I was about thirteen years old and watched *Alien* (1979) for the first time with my sister and friend, both eleven at the time. We braved most of the movie well until the first time we meet the chest burster exploding out of John Hurt, terrifying not just the cast, but us.[1] I watched through my fingers clamped tightly to my face and explained what I was watching to my sister and my friend who refused to watch.

---

[1] For those unfamiliar with the movie *Alien* and this particular story, the on scene actors had no idea what was going to come out of John Hurt's chest. The script was purposely vague and they were not given any instructions or allowed to see the set up. Director Ridley Scott wanted to see their true reactions to that scene. It's cinema gold.

What stuck with me after watching *Alien* wasn't just a new and potent fear of the unknown space surrounding us beyond Earth's atmosphere. Sigourney Weaver's Ellen Ripley dominated that movie for me. Here she was, a *woman*, taking control and giving orders and surviving until the very end and being the hero! What in the teenaged brain confusion was happening? Up until this point, every movie and show I had watched had shown women as playing second fiddle to the men, even in movies that were supposed to be about them.

Ellen Ripley became my queen, the measuring stick I used to judge women in every other movie: did she take command? Was she brave? Could she face down the thing terrorizing her and others? Was she a problem solver and did she contribute to solving the problems? Did these other women in movies give me someone I wanted to emulate as much as Ripley in *Alien* and *Aliens* (1986)? I love Ellen Ripley even through the less than stellar *Aliens³* (1992) and the mostly horrible *Alien: Resurrection* (1997). Ellen Ripley was who I judged Noomi Rapace's character against in *Prometheus* (2012), and sadly found her lacking.

I hadn't yet connected the fact that *Alien* was a horror movie, thinking that it was just a scary science-fiction movie, since, as I previously mentioned, horror movies were not allowed because of the naked people. But I was hooked. I wanted scary movies with female leads because those women could take a beating and then triumph in the end.

Shortly after that experience, Hollywood developed an obsession with remaking Japanese horror films: *The Ring* (2002); *The Grudge* (2004); *Dark Water* (2005); *Pulse* (2006); *One Missed*

Call (2008); *The Eye* (2008); *Shutter* (2008); *The Uninvited* (2009); and the numerous sequels and lesser known movies that saturated the first decade of the 2000s. Perhaps Americans wanted to see another culture's terrors after the national tragedy of 9/11. Whatever the reason, here again were primarily female protagonists fighting back against paranormal entities and being more or less successful. These women were kicking ass, and even though they weren't always successful at vanquishing the evil, I loved it. It wasn't necessarily a happy ending that I wanted; what I wanted most were women to look up to.

Given that my introduction to horror had centered around what I perceived to be a number of fierce female characters, it surprised me later in my horror movie watching career that horror was frequently perceived as misogynistic garbage which did nothing to further women or feminist ideals. These critics claimed that horror movies degraded women, ruined them in ways that made their "triumphant" endings not so triumphant. They claimed that horror was the worst genre at treating women as sex objects and punishing non-virginal females more harshly than their male partners. They howled that touting such a perverse genre as horror as benefiting women was a travesty.

And yet, for the most part, I didn't see it. Where else was I seeing women in the role of rescuer or survivor where she didn't need the help of a man? Really nowhere. That's not to say that there aren't problem genres or problem movies. The rape-revenge subgenre is frequently referenced in the misogynistic trash context and contains such delightful titles and their multiple sequels such as *I Spit on Your Grave* (1978 and 2010)*, The Last House on the Left*

(1972 and 2009), *Lipstick* (1976), and *Ms. 45* (1981). A quick search of "rape-revenge films" reveals via the Wikipedia page "Category: Rape and Revenge Films" that there are 118 American films, 15 British films, 3 French films, and 77 Indian films in that category. Now, putting aside the fact that there are more rape-revenge films coming out of America than out of India which historically and presently has an issue with keeping women safe from sexual assault and the general acceptance of it as a way of life (Parthak, S. & Frayer, L. (2019) and Krishnan (2023)), that is a lot of films dealing with the violent sexual assault of women and their supposed triumph in the aftermath.

I will note for the sake of accuracy that there are a few rape-revenge films where men are the targets, such as *Deliverance* (1972) even though the violations against the men are far less graphic than what happens to the women in movies that came out during the same time period. I will also note that some of the movies on the Wikipedia page clearly don't belong on that list, such as *The Descent* (2007) though there is a very clear revenge theme towards the end of that movie.

The rape-revenge subgenre aside, I had never looked critically at other horror movies and thought that women were treated unfairly, more violently, or more disrespectfully than men. Was I just looking at the genre through my Ellen Ripley rose colored glasses? Was there something I was missing?

Well, in fact, there was actually a lot that I had been missing and a lot I was sort of in denial about when it came to my favorite movie genre. I sought to explore that here, plumbing the depths of horror history, my feelings, and the thoughts of others on this subject. What results is, optimistically, a well thought out critical love note to horror. My hope is that you will read this with your own critical love note eyes for the genre. Just please don't pull an Annie Wilkes and try to get me to change anything.

The Tangents

Is the shark in *Jaws* male or female?

"You're gonna need a bigger boat."
Chief Brody, *Jaws* (1975)

Have you noticed that we take for granted that the shark in *Jaws* is male, despite the fact that it is never explicitly stated in the movie that the shark is male? Peter Benchley's 1974 novel apparently states that the "fish" is male,[2] but repeatedly refers to the shark as a gender neutral "it". However, the sharks in the movies and the book are all oversized fish, larger than anything anyone has ever encountered before or has probably encountered since.

In the wild, female Great White Sharks grow to be up to twenty feet long while the males typically top out at thirteen feet long. Bruce, the fictionalized shark in *Jaws* was twenty-five feet long, only five feet longer than the longest recorded female Great White[3]. So, from a scientific point of view, the shark in *Jaws* would

---

[2] I've read the book, but I couldn't find this in it.
[3] Shameless fangirl plug for Deep Blue, one of the biggest Great White Sharks on record.

more appropriately be referred to as female. And, oddly enough, fans have actually dubbed the shark in *Jaws 2* "Brucette" as the novelization of the *Jaws 2* (1978) movie states that this shark is the mate of Bruce from the first movie.

If Bruce the shark is a genderless "it" then the movie maintains its strict man vs. nature monster movie subgenre: man invades nature and nature bites back. If, however, the shark is male, does that create a deeper sort of conflict between the main characters and the shark? A modern man versus primal man conflict would lend itself to a deeper conflict than just man versus nature. The men are no longer just fighting against some monstrous beast, but against the primal versions of themselves who once existed only to eat and reproduce. We can't control nature, but we can try to control the primal urges that still reside in our crocodile brains.

If, however, the shark is a female then there is now the addition of the "monstrous feminine", a term first defined by Barbara Creed[4] to mean a feminine presence that inspires fear due to its "otherness" from the typical male characters present in horror movies. The shark has gaping jaws that shred men and swallow them whole, possibly referencing the terrifying folklore of vagina dentata (see the movie *Teeth* (2007) for this one). While I don't necessarily buy that, it is notable that strong female characters or female monsters seem to inspire a special kind of fear in men who want to deny that arbitrary female rage exists. When the shark is

---

[4] Barbara creed wrote the book "The Monstrous Feminine" which explained that men fear women because women's bodies are different from men's bodies, and that whenever women are portrayed as monsters in film, their monstrosity comes from association with reproductive bodily functions or mothering abilities.

made into a clearly female presence, as in *Jaws 3-D* (1983) (and I say this is clear because her pup dies early on as a captive in an aquarium), the motivations of the monster change from some sort of undefined and arbitrary rage against humanity, to a mother's rage at her child being taken from her and killed. The mother's rage motive seems much more trite and stereotypical, much less scary, than a large female predator who eats people just because she can.

I asked my friends on Instagram what they thought of the shark with a very scientific poll:

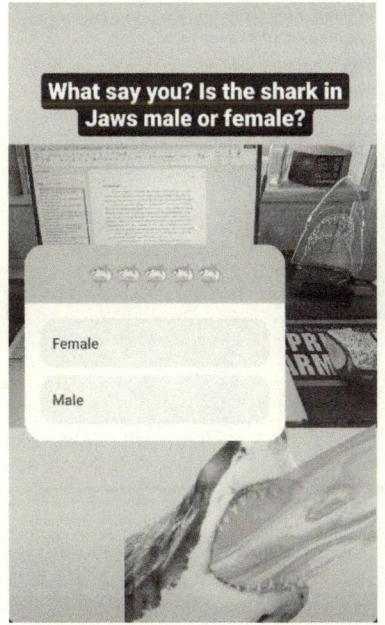
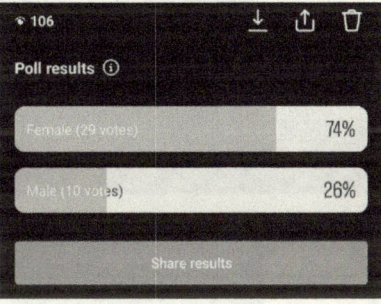

Clearly, it's not a very expansive poll with only 39 participants, but there is a clear bias in how people feel about whether the shark is male or female. Three people chose to expand on their choice:

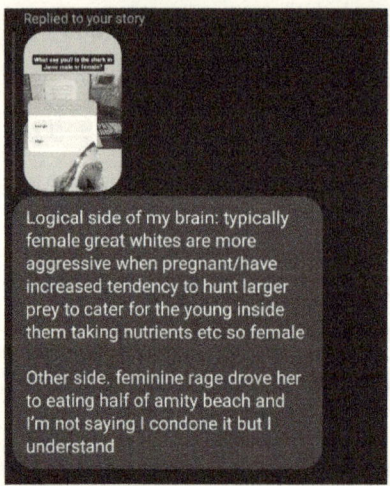

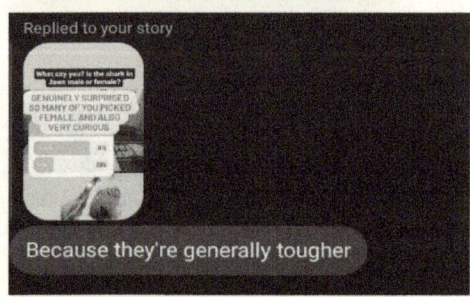

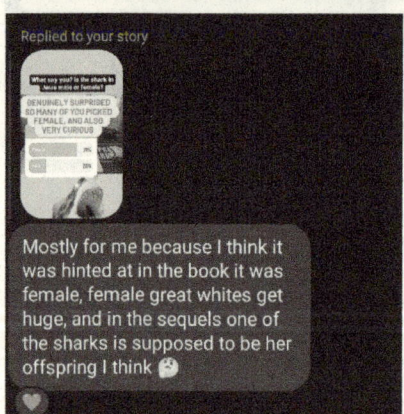

At least for these three respondents, they appear to be following along with the ideas proposed above: using science, allowing for feminine rage, and because of what was written into the books and the sequel movies.

In the end, whether male or female, the humans win out against the shark. Whether you want to read that as man conquering nature, his own primal urges, or women, it's never the swarthy man's man who makes it to the end. It's the middle-class man (Brody), the effeminate male (Hooper), or female (Ellen Brody) who triumphs in the end.

So, does it even matter what gender the shark is? No, not really. But I really like the idea that the very first horror movie I watched, in fact the very first summer blockbuster, presented a terrifying female antagonist that, to this day, keeps people out of the water and strikes fear into their hearts.

What is Horror Anyway?

"Horror films don't create fear. They release it."
                                                        -Wes Craven

What do you think of when you think of horror? Monsters, ghosts, dark basements, big stabbing knives? How does something get classified as horror? Is it the way it makes you feel? Does it have to inspire terror? Or do you know it when you see it, as Supreme Court Justice Potter Stewart famously said in regard to pornography (*Jacobellis v. Ohio,* 1964)? What is horror anyway?

The word "horror" comes from the Latin *horrere* which means to tremble, or shudder. Merriam-Webster Dictionary defines horror as "painful and intense fear, dread, or dismay"; "intense aversion or repugnance"; "the quality of inspiring horror: repulsive, horrible, or dismal quality or character"; "something that inspires horror"; "a state of extreme depression or apprehension"; or as an adjective, "calculated to inspire feelings of dread or horror".

That's a lot of words and nitpicking for something that people usually define, much more simply, as something that scares them. Although Merriam-Webster does not define "horror movie", MacMillian Dictionary did with the fewest words possible: "a movie intended to frighten people". We get frightened through suspense or gore or jump scares or things that play to our deepest fears. Home invasion movies work so well because who hasn't at some point thought about the possibility of someone breaking into their home?[5] Paranormal or supernatural horror movies work because we fear the unknown and things beyond our control. Slashers, the old reliable of the horror genre, work because they're gory and the killer doesn't stop coming until he is finally vanquished, usually by the Final Girl.

It seems that a horror movie, quite simply, could be anything meant to play to a specific fear or a multitude of fears. But the genre is not that simple.

Have you ever seen a horror movie and for the first time realized you had a very specific fear of something that arose only because you watched that movie? Take for instance *Jaws* (1975) and all the movies it spawned. Many people probably didn't realize how incredibly terrifying sharks could be until they saw that movie. Or the *Hostel* (2005-2011) film series. It is extremely unlikely (though not impossible) that American tourists are routinely snatched on vacation and sold off to be tortured to death.[6] Yet it still spawned harsh criticism from the Slovakian MP Tomas Galbavy who called the film "a monstrosity that does not at all reflect reality" and which damages "the good reputation of Slovakia" (BBC News, 2006).

These are things that, ordinarily, you may never have even thought of. Yet there they are, suddenly spawning in your brain because a movie brought up that possibility. And then you discover

---

[5] I think this fear is best realized in *The Strangers* (2008) when Liv Tyler asks, "Why are you doing this?" and Doll Face responds "Because you were home."

[6] See The Tangents 2: What Are Your Odds of Dying in a Hostel?

that there are almost as many horror subgenres (and sub-subgenres) as there are phobias: gore, psychological, monster, paranormal, home invasion, extreme, supernatural, meta, comedy, and the list goes on. You might be asking at this point, *sub*-subgenres?! Yes, there are sub-subgenres of horror.

Take for instance the Monster subgenre. Easy, you might say, only a few sub-subgenres here. Vampires, werewolves, zombies, and other lesser known creatures can be lumped together. Well, no, there's more. Are the zombies in the movie your typical undead brain chompers, or are they fueled by some sort of virus? Vampires and werewolves are a little easier since there aren't too many specific variations on them: their rules have remained hard and fast for the most part. But then you have your classic monsters (think Frankenstein or the Mummy) and your neo-monsters (Pumpkinhead). Animals (like *Jaws* or *The Birds* (1963)) are their own monster category. Giant monsters (the 50 foot woman or the Cloverfield monster) are separate from your sometimes cute and usually bad small monsters like Gremlins or Critters. Monsters can also encompass aliens or your backwoods rednecks (*The Texas Chainsaw Massacre* (1974) or *The Hills Have Eyes* (1977)) depending on how you look at it. It's possible to go on, but I think I've made the point. Horror as a genre has spawned dozens of terrifying little babies that have each developed their own roots to dig into the fear centers of our brains.

Even people who may find horror movies amusing, have that one thing that they will not mess with in horror movies or real life. In a multiple-choice survey question[7] of "Do horror movies frighten you?", one respondent selected the option of "I find them amusing.". However, this respondent also has a great fear of mirrors and gore, and avoids watching movies or movie scenes where mirrors and/or gore play a prominent role.

The monsters or fears in horror may not be the obvious things either: the creatures under the bed and in the closet, or the savage animal lurking around your property waiting for you to come out at night. Fears in horror films can also be the socially constructed kinds, where beneath the monsters and gore is a commentary on the fears we have as a society: losing a home; losing bodily autonomy; losing your job or being useless; or even, perhaps, the fear of female power. Some researchers, like Carol Clover, have postulated that the torture, sexualization, and subjugation of women in horror movies is due to men's fear that women will take their power from them. This was a very real fear in the 1960's with the advent of feminism, the Women's Rights Movement, and more women entering the workplace. There are few things more terrifying to people in power than the idea of losing that power. Film, being primarily male led and driven, would be an easy outlet for expressing this fear.

So, what is horror? How do we decide what is and is not a horror movie? I would propose that it is an amalgamation of all the things already discussed. It is something meant to scare you on multiple levels, whether superficially through jump scares or more deeply through psychological terrors. It may prey on your sense of decency through explicit and graphic scenes of torture and gore

---

[7] The survey is one which I created and disseminated so it's not entirely scientific. The full survey can be found in the references.

where the only purpose behind the suffering is to also make the viewer suffer and squirm in return. It is something that might make you think twice before entering a dark room or peeking under your bed after you've heard a weird noise.

Dawn Keetley from the website "Horror Homeroom" suggests that for something to be considered a horror movie there must be something monstrous and there must be a lack of choice on behalf of the actors. She comes to this conclusion after considering the differences between *Everest* (2015) and *Frozen* (2010). Both movies depict events happening to groups of people stuck on mountains: *Everest* is about the disastrous mountain climbing accident in 1996 on Mount Everest[8] and *Frozen* is about a group of skiers who get attacked on a mountain by a pack of wolves (not to be confused with the Disney movie).

Keetley notes that although *Everest* invokes feelings of fear and dread in the viewer by showing them harrowing incidents and traumatic injuries which might be right at home in a horror film, it does not have a defined monster and the climbers never had a lack of choice. The mountain and the storm that surrounds it which kills the climbers is what it is: a natural, though gargantuan and awe-inspiring geological formation accosted by weather events that are regularly severe. Though the storm might have been a once in a lifetime event, it is also not to be unexpected for the mountain, so therefore neither can it be considered monstrous. What may seem monstrous to the men is totally normal for the mountain. The climbers also

---

[8] Novelist Jon Krakauer participated in this climb of Mt. Everest when this disaster happened. He wrote an excellent book about the experience called *Into Thin Air*.

made the choice to be there, knowingly accepting the risks for a potential reward.

Frozen, on the other hand, has an identifiable monster: the wolves. The movie takes place in New England which is decidedly lacking in wolves, to the point where one character asks when the last time anyone had heard of a wolf attack in New England. Real life wolves typically avoid humans at all costs, yet the wolves in *Frozen* seek out the marooned skiers in order to tear them apart. They even chase down and kill one of the experienced skiers as he's trying to escape downhill. Something natural has been made unnatural and is menacing unsuspecting victims, something which happens in many horror movies. As for choice, the skiers made the decision to go skiing, however getting stuck on a ski lift and being subjected to marauding wolves is not typically an accepted risk of the sport.

The requirements of lack of choice and a monstrous force to help delineate horror from non-horror in what can sometimes be a very thin line are helpful tools for the definition. A rather famous example of what many people consider a horror movie, but more high-brow audiences consider a thriller, is *The Silence of the Lambs* (1991). It swept the Academy Awards, being only the third film in history to win Best Picture, Best Director, Best Actor, Best Actress, and Best Adapted Screenplay. It did all this while being considered a thriller, and not the horror movie that it really is. Cynically, I would say this is because horror is looked down on as a genre. It is thought of as being cheaply made with no plot and just for watching teenagers get hacked up on screen. Horror movies don't win awards because they're not good enough for polite society.

The Silence of the Lambs, though, was too good even for polite society to overlook. So they rebranded it a thriller and threw awards at it. It is undeniable that the movie has thriller elements to it: the police procedural, crime investigation, new cop (or FBI agent) thrown into a tough situation. But *Lambs* also has the monstrous. Hannibal Lecter is not your run of the mill serial killer (who is still present in regular horror movies), but a serial killer cannibal who came from money and education. He is monstrous not just because he kills people, but because he eats them, and he does it with intelligence and menace. He even manages to achieve some preternatural abilities as well, inexplicably managing to escape his cell and pretending to be a dead cop, later eluding the FBI. He is something natural – a human – made unnatural through his violence, cannibalism, and his inhuman abilities to connect with people and get what he wants.

The lack of choice is also present in the film. Clarice Starling is given the position of interviewing Lecter, and the victims of Buffalo Bill do not make the choice to be kidnapped and skinned to be worn as a body suit for the murderer. Lecter is constantly manipulating everyone around him to achieve goals that only he is aware of, and the others can only grasp at. *Lambs* definitely fulfills both of those qualifiers.

Then if you look at the atmosphere of the movie, it becomes even more undeniable. Lecter is not held in a normal prison, or even a high security prison, but in the basement of a psychiatric facility that is reminiscent of dungeons in medieval castles. This basement dungeon is filled with monstrous others, rejects of society who do not abide by social norms, one of whom throws his ejaculate into

29

Starling's face. This is not your typical thriller fare. Even without the gore of other horror movies, the implied violence of being skinned and worn as a skin suit hits nerves and fears that run deep within the horror genre and not typically seen in thrillers.

Maybe viewers, academics, and award grantors would feel better about horror if it was more "elevated". Personally, I don't like the elevated horror subgenre. When directors set out to make elevated horror, it is as though they want to make the most artsy, sort of violent, complicated movie that they possibly can. Movies like *Midsommar* (2019), *The Witch* (2015), *The Lighthouse* (2019), and *Hereditary* (2018) are all considered to be elevated horror films. They are slow paced, focused on relationships and current social issues, using horror as a vehicle to convey a message rather than making terror the message. Somehow elevated horror wants to separate itself from "low brow" horror like slashers. The problem with that, though, is that these low brow slashers and other horror subgenres can also convey social issues. Sometimes, the impact is even greater when people are able to root out their own social issues and meanings rather than being smacked over the head with the director's perception of what should be said about social issues.[9]

Hereditary is considered to be an elevated horror, combining commentary on generational trauma with a demonic cult element.

---

[9] This might be seen most clearly when comparing Rod Serling's *The Twilight Zone* to Jordan Peele's *The Twilight Zone*. Serling's series lasted over a decade with 150+ episodes. It has a 9.1 rating on IMDB and an 84% on Rotten Tomatoes. Peele's reboot made it through only two ten episode seasons and carries a 5.9 and a 66% on IMDB and Rotten Tomatoes respectively. I firmly believe this is because Serling's *The Twilight Zone* shared it's political and social commentary smartly and subtly. Peele's *The Twilight Zone* tried to make its points with a sledgehammer.

The movie is long, packed full of quiet moments that are supposed to be contemplative and tense, evoking a sense of fear and confusion. For the casual horror enjoyer, the movie is too long with too many themes. The ending is confusing and unsatisfying. It had been trying to build up to this terrifying demonic cult and possession story, but it gets lost in the annoying antics of Charlie, the histrionics of the mother, and the weird acquiescence of the husband.

On the other hand, *Umma* (2022) tackles not only generational trauma, but immigrant trauma. These stories are woven in with the supernatural ghost story of Amanda being haunted by her mother's ghost. The movie is spooky with a clear storyline and clear reasoning behind it. There are no long scenes of silence to create tension; the tension creates itself between Amanda's struggles with her past and with trying to maintain her relationship with her daughter, Chrissy. Things hide in the dark corners, not just to create a jump scare, but to make you question what you are really seeing. There is some guesswork, but the movie does not draw it out.

What makes Ari Aster's *Hereditary* an elevated horror but Iris Shim's *Umma* psychological/supernatural horror? Is it because *Hereditary* was done by man with a big name in Hollywood and *Umma* was done by a less well-known woman?[10] I couldn't say.

The Texas Chainsaw Massacre (1974) is an example of slasher horror with a message, albeit one muddled by the disturbing nature of the film. Tobe Hooper stated that the movie was supposed to be a commentary on job loss and the difficulty of blue-collar

---

[10] I really hope this view doesn't come across as too hateful. Although my dislike for elevated horror is clear, I hope the points I'm making are also clear and objective.

workers trying to find work in a changing world. The hitchhiking brother the teens pick up in the 1974 *Texas* makes this very clear when he talks about how his family lost their jobs when the cows stopped being slaughtered with hammers. Perhaps the message is lost in the initial shock of the viewer hearing that this man would prefer cows be slaughtered less humanely if it meant his family still had jobs. For people who did pay attention, who wouldn't choose one method over another to make sure their family could live and survive? This message carries across all of the *Texas* movies and continues to be relevant despite the decades long gaps between movies. People will always be afraid that automation and advances will take their jobs. *Texas* showcases real life horror mixed in between the horror of getting murdered by crazy men in masks without being blunt about it.

George Romero's titular zombie film *Night of Living Dead* (1968) is heralded by many as being a social commentary on race issues in America. Ben, acted by Duane Jones, is the lone black character in the film, surviving until the end only to be shot by police who mistake him for a zombie while he's calling for help. This seems pretty intentional for the time period. Romero, however, has said that he didn't intend for the movie to be a commentary on race. Duane Jones just happened to have the best audition for the role. Regardless of Romero's intent, I have never seen *Night of the Living Dead* referred to as an elevated horror. While I would argue that it's important for a genre to continue to evolve and change to follow the times, sometimes those evolutions don't always lead to viable offspring.

Horror is, and can be, what we make of it within a variety of definitions. It is as broad and deep as the number of subgenres (and sub-subgenres) that it has spawned. We can stick to the hard and fast definition, or we can let the gray areas which absolutely exist in the genre bleed through. Horror can, and does, bleed through not just into different kinds of movies, but into aspects of our lives. All of our experiences are subjective to who we are and what we've experienced.

The Tangents

What are Your Odds of Dying in a Hostel?

"You won't find this hostel in any guidebook."
-Alexi, *Hostel* (2005)

Horror movies tap into things that we are afraid of, plucking on the strings tied into our anxieties and our fears, those on the surface and the deeper ones that we haven't yet discovered. Most people probably thought nothing of staying in hotels, motels, or hostels so long as the places were clean and the doors had locks on them. The *Hostel* series (2005-2011) brought about a fear of staying in these places, especially overseas, to the forefront of everyone's minds. Oddly enough, Airbnb started up three years later in 2008 and is still doing well despite some extremely creepy and scary incidents, so maybe people weren't as affected by those movies as the Slovakian government thought they might be.

The thing is, we can all point to a movie that has made us question whether or not we should do something or partake in some action. Swimming in the ocean? Having an open house? Playing with a Ouija board? Spending time by yourself out in the country?

Okay, playing with a Ouija board should always be second guessed, but the other things seem pretty run of the mill. Yet people still avoid the water or graveyards or being alone in unfamiliar places, things that are far less likely to kill you than many other things we do every day.

If you're still wondering if your odds of dying in a hostel are greater than, say, your odds of dying in a car crash, which are 1 in 93 (National Safety Council, 2021), you're not in luck. There doesn't seem to be much data about it, although one study done in Copenhagen, Denmark found that over a ten-year period, young homeless women were more likely to be killed in hostels than others in the general population (Nordentoft, Merete and Wandall-Holm, Nina, 2003). Other than that study, there's not much detailing death or kidnap trends from places people stay in when they're away from home. That could mean that it doesn't occur that frequently, or it could be because statistics are difficult to come by concerning things like kidnapping for human trafficking.

When you consider that Eli Roth based his movie off a website from Thailand that offered to let you kill a person who supposedly volunteered to be killed in order to get money to feed their family, the series becomes a bit more insidious. Roth was never able to confirm much as proceeding further onto the website required a credit card, but the fact the website existed at all is concerning. The website from Thailand that would provide the thrill of murdering someone for money is not farfetched at all when you learn that you can access the Dark Web through a generic website and have access to an assortment of weapons, credit card numbers, body parts, and people for selling into slavery.

It is doubtful that people think much about *Hostel* anymore, or let it influence whether or not they travel or stay in certain places. Or maybe they do. I still avoid the ocean over a 1 in 3,748,067 chance that I'll get attacked by a shark (McLaughlin, K., 2023). But you can bet the house I'll pet a strange dog even with a 1 in 53,843 chance of being killed by a dog attack (National Safety Council, 2021).

Fear and avoidance are really weird things.

A (Brief) History of Horror

"What scares me is what scares you. We're all afraid of the same things. That's why horror is such a powerful genre."

-John Carpenter

Now that we've got a basic understanding of the definition of horror and what generally goes into it, we can talk about women in horror. Right?

Well, not entirely. Although horror films are only a little over a century old, they have a long and rich and very tangled history. Jumping straight into asking "Where do women come into this?" or "What roles have women historically held in horror movies?" skips over a lot of information relevant to understanding the genre. A good question to start with could be "Was there even a woman to be found in the first horror movie?" Men frequently played the roles of women in theater and films during the times before women were allowed to do anything other than take care of the home and have babies. Greeks and Romans felt that theatre was "dangerous" for women, and the Christian Church in Europe felt that it was improper for women to be in theater. Definitely no Final Girls or even sacrificial whore girls to be found here.

Early horror focused on folklore and the supernatural, forgoing blood and gore, and instead frightening people by showing them weird and fantastical things on screen. Although, I suppose at that time seeing anything on the screen was a fantastical experience. I wonder how magical it must have been to see movies when they were so incredibly new, when movies weren't taken for granted and no one lamented poorly done CGI and special effects. Early horror was able to take advantage of the newness of movies and really focus on frightening people.

The first recognized horror movie, which used "new" special effects and supernatural folklore, was done by George Méliès in 1896 called "Le Manoir du Diable", also called "The Haunted Castle" or "The House of the Devil". It is a three-minute-long film and is watchable on YouTube with a variety of scores as Méliès would opt for theatres to put their own scores over his films. It is, of course, in black and white and has no dialogue or title cards to provide dialogue. The film features several spooky items like a bat, a cauldron, a skeleton, and witches complete with brooms, casting some sort of spell. I find it interesting to look back at very old films (this one is one hundred and twenty-four years old!) and see what would scare the people from that decade. "The House of the Devil" is unlikely to scare even small children now, but back in 1896 I'm sure there were at least a few people spooked by cutting edge special effects for that time period.

The majority of the characters in this three-minute film are women, and they also appear to be taking on the role of villain as the witches. The male protagonist who starts doing all the magic even conjures a woman from his cauldron who turns out to be a witch.

Kimberly Pinzon

The reasons for the witches' behavior and how they are vanquished aren't clear but the fact that women are the villains in the first horror film is pretty awesome.

Méliès then came out with "La Caverne Maudite" or "The Cave of Demons" in 1898. In this short film a woman finds a cave filled with the spirits and skeletons of people who died there. He has another film that focuses on women called "Le Papillon Fantastique" (1909) where a magician summons a butterfly woman, and then a spider woman who attempts to eat the butterfly. The magician manages to save the butterfly woman (perhaps the first Final Girl, if one were being generous) and vanquishes the hungry spider woman. It is at this point in time that horror movie films become more prominent, particularly those where film makers use novels as the source material for their films.

Literature was, for the most part, more advanced than film in its inclusion of female characters, so movies based on books would necessitate the addition of woman characters to the films. This might be why you see women in more films at this point, or it could be that it was much harder to make a film more than a few minutes long if you did not have enough characters to build a story around or interact with each other. Regardless, we roll through this Literary Decade with film introductions to Dr. Jekyll and Mr. Hyde, Frankenstein, and vampire bats in the 1912 French film serial "Les Vampires".

Frankenstein (as the monster and also Dr. Frankenstein, the creator), a cultural mainstay in film since his debut, was written by a woman named Mary Shelley. As we move through the next few decades, you will notice that some of the most memorable and classic horror films were done by men based off stories written by women. Unlike with Frankenstein though, where Mary Shelley's name is inextricably linked with the monster and the doctor who created him, these women, like Daphne de Maurier[11], are largely forgotten and given no credit for the stories they created.

Since George Méliès was introduced as the first male director of horror, we need to talk about Alice Guy-Blaché, the first female horror director who has been largely written out of her own history. Not only that, Guy-Blaché pioneered filmmaking for stories with narratives, rather than just showing panoramas of wildlife, nature, and people as the influential Lumière brothers had done. *The Cabbage Fairy* (1896) is the first narrative movie, showing babies being born in cabbage patches with fairies. Many of her films focused on the double standards of gender norms and her film *A Fool and His Money* (1912) is considered to be the first narrative film with an all-black cast. Her first horror film was adapted from Edgar Allen Poe's "The Pit and the Pendulum" in 1913, during the decade when many directors were starting to delve into the horror genre and were adapting literature to movies. In her movies she developed techniques such as double exposure which still appear in horror movies today. She mentored Louis Feuillade and Lois Weber[12] who made important contributions to early horror filmmaking, and Alfred Hitchcock cited her as inspiration for his work. Yet Alice does not

---

[11] More on her a little later.

[12] Weber was a prolific silent film director, becoming the highest paid studio director – man or woman – in 1916. She is referred to as "the most important female director" of the American Film Industry and "the most important and prolific film director in the era of silent film". She featured nudity, prostitution, birth control, and other taboo subjects in her films to highlight social issues.

appear in the history of The Solax Company for which she was Head of Production, or in books from that time period. She has suffered the fate of many women who made contributions to a man's world: being forgotten. It would probably come as a surprise to many that 50% of silent era films were written by women. During this time Universal Studios employed eleven female directors who made more than 170 films (Bose, 2021). It seems that once films started to become profitable, men restructured Hollywood to benefit themselves and push out women.

Following this time period, the 1920s saw the start of The Golden Age of Horror, where movie makers wanted to focus on scaring people rather than just providing dark melodrama and gothic horror themes. *The Cabinet of Dr. Caligari* (1920) and *Nosferatu* (1922) are inarguable classics from this time period, with *Nosferatu* cementing many of the vampire cliches.

The first installments of Universal Picture's monster films started with an emphasis on deformed characters like *The Hunchback of Notre Dame* (1923), *The Phantom of the Opera* (1925), and *The Man Who Laughs* (1928). A haunted house style movie called *The Cat and the Canary* (1927) directed by German Paul Leni and a 1929 film called *The Last Performance* rounded out Universal's horror stock for the 1920s. If you are wondering how 1923's *The Hunchback of Notre Dame* could possibly be a horror movie, you should get rid of all of the notions you have about the story from the Disney version and watch this silent black and white version. It will all make sense. What is notable here is that Universal is starting to dip its metaphorical toes into what it will repeatedly attempt to create into a universe and repeatedly fail at doing so (looking at you *The Mummy* from 2017).[13]

The 1930s see movies being described as "horror" for the first time. Prior movies were considered to be dark melodramas with emotion driven plots that had gothic horror themes. The 1930s also saw the entrance of horror's first stars coming to the screen such as Bela Lugosi who was pretty much the first mainstream Dracula and appeared in other great horror movies like *White Zombie* (1932) and *Abbott and Costello Meet Frankenstein* (1948)[14]. The more influential Frankenstein movies, *Frankenstein* (1932) and *Bride of Frankenstein* (1935) were released during this decade. Although previous directors had pushed the boundaries of what was acceptable in film, the proliferation of film brought it into the public eye which also brought the potential for controversy. *Freaks* (1932) was so shocking at the time that it was cut extensively and ruined the career of director Tod Browning. It was banned in some countries and is still considered to be very disturbing even now, almost a century later.[15]

The forties were a quiet decade for horror likely stemming from the fact that America was heavily involved in World War II for the first half of the decade. It did feature many versions of the Wolf

---

[13] Nothing will ever beat *The Mummy* from 1999 and Universal should really stop trying.

[14] I originally included *Plan 9 From Outer Space (1959)* here, but, I thought that might be a bit generous regardless of how much I like it.

[15] The movie is genuinely disturbing and could easily be mistaken for a movie done in the present day, just in black in white. The disabled performers in the film are all actually disabled. None of what is seen is done with the use of prosthetics or movie "magic" except for what happened to Cleopatra at the end. Probably.

Man, Frankenstein, the Mummy, and the Phantom of the Opera. There were a few more notable films such as *The Monkey's Paw* (1948), *The Ghost and Mrs. Muir* (1947), and *Gaslight* (1944). *Gaslight* is especially pertinent because it spawned the term which would continue to be prevalent when talking about emotionally and mentally abusive situations and people trying to trick others into disbelieving their circumstances[16]. It's also a common theme in horror films where protagonists are made to think they are crazy when they see evil or supernatural entities.

Fears of nuclear fallout and invasion stemming from World War II, the atomic bombs, and the ensuing Cold War spawned movies dealing with radioactive mutations, monster invasions, and aliens in the 1950s. These include such memorable classics as *Invasion of the Body Snatchers* (1956), *The Blob* (1958), and *Attack of the Crab Monsters* (1957). One of Japan's important contributions at this time was Godzilla, first appearing in 1956. The giant radioactive lizard who is, depending on what movie you're watching, both lovable and terrifying, became one of the most permanent fixtures of fear relating to nuclear war and fallout.

The 1950s also saw horror's attempts to get interactive with the audience. William Castle's *The Tingler* (1959) is a great movie that hooked up seats in the theatre to buzz movie goers when the tingler escaped captivity. There were also actors paid to faint in the theater, and Vincent Price breaks the fourth wall and interacts directly with the audience when the screen goes black and he tells

---

[16] In the film, the husband changes the intensity of the gas lamps throughout the house and tells his wife that she is imagining things when she comments on it. Thus the term, "gaslighting".

viewers not to panic, everything is under control. People absolutely still panicked. I definitely would have.

The 1960s transitioned into a more serious tone for horror movies. George Romero's 1968 classic *Night of the Living Dead* is credited as kick starting zombie movies and starred a black man in the lead role in a movie that wasn't about race. *Rosemary's Baby* (1968) showed the terrors of pregnancy, gaslighting, and satanism. Two of Alfred Hitchcock's most famous movies, *Psycho* (1960) and *The Birds* (1963), made their debuts. Following this influential decade of psychological and supernatural terror, the 1970's and the 1980's ushered in and established the slasher and occult film subgenres. *The Exorcist* (1973) and *The Omen* (1976) are the most recognizable, but there are many movies in the 1970's which reference Satan, the devil, the bad seed, or demons.

The slasher movie got its start in the 1980's with a slew of movies involving murdering teenagers. These movies also inspired multiple sequels, prequels, and reboots, becoming some of the most recognizable faces of horror. Even if you have never seen the movies, you are undoubtedly familiar with *The Texas Chainsaw Massacre* (1974), *Black Christmas* (1974), *Halloween* (1978), *Friday the 13th* (1980), and *A Nightmare on Elm Street* (1984). The success and staying power of these movies unfortunately led to a reliance on formulaic writing which often performed worse and worse with each new installment. Monster movies like *Jaws* (1975) and *Alien* (1979) also suffered from this trend, putting out excellent sequels followed by two less than stellar additions.

Horror supposedly lost some of its steam in the 1990's, with critics and researchers referring to this decade as The Doldrums.

Audiences were tired of the slasher sequel saturated market where one movie looked like the next one. They were predictable with their teenage cliché characters, "unstoppable" murdering monster, and cookie cutter endings. I've watched more recent movies and made the comment, "This is like a generic 90's horror movie" because of all the horror stereotypes that were reinforced in that decade. Studios weren't making an effort to put out anything new. Or were they?

What about memorable films like *Tremors* (1990), *Misery* (1990), *The Silence of the Lambs* (1991), and *Arachnophobia* (1990)? But wait, there's more! Because this decade, though frequently thought to be a slow decade, also had some of the movies which impacted the horror genre even up until present day. *The Blair Witch Project* (1999) became the most wildly successful film ever based on how much it cost to make versus how much it earned, and breathed new life into found footage films[17], showing that a good story could be done even on a small budget. *The Sixth Sense* (1999) launched M. Night Shyamalan's career and provided one of the best movie twists ever. And I mean that sincerely. This book is rife with spoilers but this is one movie that I will not spoil because it's just too good not to discover on your own. Wes Craven's *Scream* (1996) mocked horror tropes by calling them out and subverting them, giving new life to the definition of Final Girls who were allowed to be people rather than just a trope. Finally, *Candyman* (1992) brought

---

[17] For an extensive found footage list, check the references for this section. Prior to *The Blair Witch Project*, there were relatively few found footage films with the most recognizable as *Cannibal Holocaust* (1980).

to life a black villain who was complex in character and motive, and created a new way for people to be frightened of mirrors.

The 2000's were a bit of a weird time for film in America. There were multiple Japanese film remakes, more reboots of old series, and the development of the torture porn subgenre. *Saw* (2004) and *Hostel* (2005) kicked off this subgenre which is largely thought to be the result of 9/11 and the hopelessness America felt as a country. There was a general idea that America had suffered so immensely that there was nothing in horror that could be shocking anymore unless there was extreme gore and senseless violence. 9/11, as I suggested earlier, might also be the reason Hollywood remade so many Japanese horror films. Japanese folklore is full of murderous ghosts and supernatural creatures. If America did not want to face any of its own demons, or any more trauma on our own shores, pulling inspiration from a country which we already had so many pop culture influences from (Pokémon, Digimon, manga, anime, Hello Kitty, Gwen Stefani's whole kawaii Harajuku Girls phase, need I go on?) made Japan an obvious choice.

There were also comedic horror movies or movies which mocked and subverted horror tropes to comedic effect. *The Cabin in the Woods* (2012), *Shaun of the Dead* (2004), *Zombieland* (2009), and the *Scary Movie* series (2000, 2001, 2003, 2006, and 2013) are all great examples of this.

Later in the 2000's and into the 2010's a new-ish genre called elevated horror started to take shape. The "new" is qualified by an "-ish" because although the term "elevated horror" was not really tagged onto movies until Robert Eggers (*The Witch* (2015), *The Lighthouse* (2019)) and A24 began to produce movies, "art horror" has been around since the 1920's. Some researchers will even cite older works such as Alfred Hitchcock's *Psycho* (1960) as being a part of the elevated horror genre. The kinds of films that are typically identified with this term are those which explore psychological themes and the human condition through allegory, metaphor, and subtext without using gore, jump scares, or other low horror elements. These movies, directors, and some of their fans perceive these films as being art, movies that you don't have to feel dirty watching because there isn't an excessive amount of gore and murder. Movies that frequently pop up in lists for the best elevated horror genre films include *Hereditary* (2018), *Midsommar* (2019), *It Follows* (2014), and *The Babadook* (2014). There is debate about whether this term should exist, if it is elitist, or if it genuinely and accurately describes a subgenre of horror.[18]

While horror is typically viewed as a genre for men, in doing research for this book, I came across many female driven films which have been largely overlooked. In a short article from the website "Horror Homeroom" which had been published in its full form in the *Journal of Film and Video* 68, no. 1 (Spring 2016), the writer, Dawn Keetley, brings up a movie that I had never heard anything about called *Thirteen Women* (1932). She posits that it is the first of the slasher subgenre, predating *Texas Chainsaw Massacre* (1974), *Black Christmas* (1974), and *Halloween* (1978) by four decades. The film's description on IMDB is "Ursula plots to murder twelve women with her supernatural powers" and if that doesn't sound like something straight out of an eighties slasher movie I don't

---

[18] See more in "The Tangents: Why are people so high on "elevated horror"?

know what you're looking for. The cast is almost entirely female and the villain is a minority woman, making her doubly out of the norm. Though Ursula, the villain, states that her primary motivation for killing the sorority sisters is because they would not let her into their sorority, she also vaguely mentions a possible rape that occurred when she was a young girl by white sailors. Acknowledging a rape, especially of a minority woman by white men would have been especially taboo. Ursula manipulates all of the men around her to assist in killing the women and, in some cases, kills the men she was using. The film also focuses repeatedly on Ursula's eyes, and the people she focuses her eyes on. The story is clearly being told from her point of view. At the end, Ursula is even accused of being "inhuman", just like many of the prominent male slasher villains. *Thirteen Women* is a fascinating look at the precursor to slasher films. I wouldn't say it is a perfect representation, but it deserves more attention that it has received.

Daphne Du Maurier, mentioned earlier, was a 1930's writer who gets unfortunately categorized as a romance writer which makes no sense when you consider that her works "Rebecca", "The Birds", "My Cousin Rachel", and "Don't Look Now" have all been the basis of classic horror films. She chose monsters that could hit closer to home through our family members and people that we love. Yet when we think of *The Birds* or *Rebecca*, it is Alfred Hitchcock we think of, and not Du Maurier.

The 1950's experienced not only the rise of film noir, the space age, and Alfred Hitchcock, but also the return of more women in the horror genre and industry. Ida Lupino contributed *The Hitchhiker*, a 1953 psychological horror in which two men pick up a

50

hitchhiker who turns out to be a serial killer who torments the men.
Gore doesn't factor into this film where the real focus is on these
men's masculinity and the fears associated with it. Lupino again
plays on our fears of who we really are on the inside with her
episode on *The Twilight Zone* called "The Masks". Ida was the first
female filmmaker to direct a film noir and to direct an episode of *The
Twilight Zone*, opening doors for other women to enter the genre
again.

It is postulated that as men began to realize that films could
be more profitable, they took more and more control over the
industry and gradually pushed women out of what has become a
"man's world". Or, that when the independent film organizations
grouped themselves together into just a handful of studios, the men
pushed the women out. Pretty much a six in one hand, half a dozen
in the other situation. Hollywood's "Golden Age" is considered to
be the late 1920's until about the early 1960's. Prior to the 1930's,
women made up about 40% of casts, wrote 20% of movies, produced
12%, and directed 5%. Once the so-called Golden Age hit, those
numbers plummeted: women made up only 20% of the casts and
were essentially absent in producing and directing roles (Morris,
2020).

Today, these numbers have improved (vastly in some roles),
though the gaps are still obvious. There are some instances, such as
with writers, where the percentage of women has actually gone down
from the pre-1930's numbers. In 2016, women held only 37% of
major roles in top grossing films (Lauzen, 2017). In 2020, 85% of
films had more male characters than female characters (Lauzen,

2021). In 2021, women made up 17% of writers, 17% of directors, and 32% of producers (Lauzen, 2022).

Horror is not the only genre struggling with its representation of women. It's just the genre that critics like to pick on.

The Tangents

Why are people so high on "elevated horror"?

"Yeah, I do like scary movies, especially the ones that don't take themselves too seriously."
-Anna Faris

Elevated horror is one of my least favorite subgenres because it seems to look down on all of the subgenres I enjoy. The films purport themselves as being psychological horrors that have important thematic elements which convey a message without using gore or jump scares. One of the words I see most frequently used in describing a film in this subgenre is "atmospheric". John Carpenter has said that he doesn't know what elevated horror means and that all movies can have a theme or a message. Jordan Peele has balked at the idea of his films being elevated horror (both *Get Out* (2017) and *Us* (2019) have featured repeatedly on elevated horror movie lists). Peele has essentially said that he does not want people to label

53

his films, or to think that he sets out to make prestigious films that will be labeled as elevated. He would rather his films be something that people perceive as being weird, or something he shouldn't have made because they're so "fucked up" (Sharf, 2022). Other directors of this genre seem to want their work to achieve prestige and notoriety based on their status as elevated horror films.

Is elevated horror really doing anything new, or better than, previous films? Many of the terms used to describe the genre include emotionally disturbing, psychologically devastating, begging interpretation, metaphorical, beautiful, experimental, and slow burn. Many of these terms could be used to describe other horror movies. *The Thing* (1982) slowly builds tension as the men become increasingly suspicious of one another and can't figure out who they can trust. The ending is ambiguous, leaving the viewer to question whether Childs or MacReady are really the creature, and knowing that it doesn't matter because the story has led them to the conclusion that they cannot trust each other anyway. Terrifying, yes. Addressing the human condition, yes. Thematic elements? Sure. But somehow this movie is not as good because it includes gore and jump scares.

Candyman (1982) would be another comparable movie. It addresses systemic racism and generational trauma not only of the film's namesake, but of the people living in the Cabrini-Green Housing Projects. This is rife with psychological and emotional strain, as well as addressing societal ills through blunt allegory. Yet the gore and the jump scares here preclude this from elevation.

Neil Marshall has explained that in his 2007 movie *The Descent* there are three distinct levels to the movie and the cave

54

system, each addressing the character's descents in grief, madness, and death. There are two separate endings to the film (one American and one U.K.), each of which can be interpreted a number of different ways. Marshall has said that there are hints hidden in the movie that the creatures are perhaps part of Sarah's imagination and she is the actual danger in the cave, having been driven to madness by loss and betrayal. Despite the blood and gore, the movie is stressful to watch with claustrophobic scenes and constant looking in the shadows for threats. It has layered meanings and interpretations, yet this has not shown up on any elevated horror lists that I have seen.

The fact that people question whether this is even a "real" subgenre or a meaningless term outside of Film Twitter (or Film X which just sounds weird) might give one pause. Is it just another form of psychological horror? The films definitely have a lot of similarities to psychological horror in how they attempt to address issues people don't typically like to look at, or how they focus on tension and character development. Though elevated horror claims that it addresses meaningful content that other horror movies don't, that is very pretentious to say when many directors have said their films have underlying meanings and horror viewers dissect and analyze horror films that are not within this subgenre.

Does the genre do more for women than other "misogynistic" and "low brow" horror movies do? I would say no. *The Witch* (2015) showcases what happens when young women are subject to unreasonable demands from their families and society regarding their sexuality, bodies, and lives. *Midsommar* (2019) shows what can happen when women are neglected by those they love (particularly their men) and find family with other communities. *The Babadook* (2014) focuses on the strains of motherhood and grief. *It Follows* (2014) is based around a sexually transmitted spirit that haunts a young woman after she is sexually assaulted by a man. All of these themes and issues have been explored in other horror movies.

What elevated horror does do, though, is attempt to bring non-horror fans into the genre. Movie viewers who don't enjoy gore or blood may be more likely to go see a film that has horror inclinations without the requisite splatter and body mutilation. Does this make the elevated horror subgenre any better than other horror subgenres? No, it doesn't. It's simply a preference for a different type of horror.

Who is The Final Girl?

"What's the point? They're all the same. Some stupid killer stalking some big breasted girl who can't act who is always running up the stairs when she should be running out the front door. It's insulting."
Sidney Prescott, *Scream* (1996)

The Final Girl trope has an almost mythic quality. She is the survivor, the victor, the one who makes it out and slays the bad guy/monster/evil entity/what have you. She is one of the most recognizable tropes and the one viewers are always looking for. Who else begins every movie trying to identify who the Final Girl will be? I sure do. The Final Girl is the one everyone wants to be: she makes it out of the horrifying and traumatic encounter as a hero. Don't mind the painful injuries and excruciating post-traumatic stress disorder she's going to suffer from, we don't need to explore that. It is unimportant because she *survived*.[19]

While Merriam-Webster defines the Final Girl as "the female protagonist who remains alive at the end of the film, after the other characters have been killed, where she is usually placed in a

---

[19] Please note the sarcasm.

position to confront the killer", this term was first coined by Carol Clover in her book *Men, Women, and Chainsaws* (1992), a collection of essays focused on gender in horror. She identifies several markers of a typical Final Girl such as being more masculine than feminine, having a boyish name (Carly, Jamie, Stevie, Max), and being a virginal good girl. The Final Girl is "the distressed female most likely to linger in memory…the one who did not die…she is abject horror personified" (*Men, Women, and Chainsaws*, pg. 35, 1992). The Final Girl is desexualized in her masculinity because female sexuality is punished. Men don't want to see their heroes be women (at least that's what Clover thought).

These women, according to the opinions of some people that I talked to (who are not researchers but average people), believe that women are chosen for the Final Girl mantle because women engender more sympathy than men in general. These average people believe that there are more extended chase scenes, longer death scenes for women, and more violence towards women because people will care more about a woman's suffering than the suffering of a man. When I suggested that including more male suffering in movies might be a more accurate reflection of real life, as men should also be encouraged to share their emotions and be seen to suffer as they do in real life, a few of those I spoke to felt that that would not be realistic.[20]

In a survey question which provided the definition of the Final Girl and then asked if she was a survivor or a victim, 55.15%

---

[20] The people who mentioned these specific opinions were men over the age of 55. It is likely that due to the time period they grew up in where men were not allowed to express emotions and had to be strong all the time, they have this perception.

said she was a survivor, 10.3% said she was a victim, and 34.6% chose the "Other" option with many of those respondents saying that the Final Girl is both survivor and victim. One respondent said she would be "forever tainted by the experience", a possible reference towards post-traumatic stress disorder, while another specifically referenced PTSD. There was an opinion that she is a "plot device. A shallow attempt to 'empower' women while simultaneously arousing men with chase scenes and heavy breathing." These opinions differ from the definitions given by Clover who viewed Final Girls as desexualized victims. But are these more realistic definitions being reflected by changing societal ideals?

When thinking of horror movies, what women would you name as a Final Girl? What traits do they have? Ellen Ripley was already mentioned, but do you recognize the names of Sally Hardesty from *The Texas Chainsaw Massacre* (1974), Laurie Strode from *Halloween* (1978), Jess Bradford in *Black Christmas* (1974), Kirsty Cotton in *Hellraiser* (1987), Sarah Carter in *The Descent* (2005), Gale Weathers and Sidney Prescott in *Scream* (1996), Nancy Thompson in *Nightmare on Elm Street* (1984)? These are some of the most classic Final Girls and their names are not boyish. They're also not particularly masculine unless you consider fighting back a solely masculine trait. As for being virginal good girls, I won't comment on whether or not the women are virgins beyond assumptions but other than Laurie Strode, they're not all exactly good girls either. Many of them have secrets or engage in behaviors that Clover would not attribute to Final Girls. Sidney and Jess are clearly sexually active. Kirsty and Nancy outsmart everyone around them. Sarah Carter has mental health and trauma issues. Clover's definition is important, and relevant in some ways, but it is not a hard and fast definition, and it seems there are more outliers than strict conformers.

Final Girls never get out of their encounters unscathed, and sometimes not even with any of their friends. They often know or suggest that something bad or dangerous is happening or will happen. They are resourceful, clever, and determined. Depending on the movie, sometimes they need the assistance of another (usually male) character to defeat the evil or the monster, and sometimes they can defeat the monster or escape on their own. They aren't "girly-girls" (though Nancy is dressed almost entirely in pink), and they don't shy away from using violence to save themselves or their friends. They are the person we would hope we could be in similar horrific situations, regardless of what gender we are as we watch the movie.

Despite my joking suggestion that the butterfly woman in "Le Papillon Fantastique" was perhaps the first Final Girl, many will immediately think of Laurie Strode as being the first woman who fulfills that role. She is a good girl with good grades who doesn't associate with boys and she's always willing to help her friends out. She runs from and fights against Michael Myers, whom she had previously noticed stalking the neighborhood and no one believed her when she mentioned his creepiness. She gets a few hits on Michael, while incurring some injuries of her own, all while finding the murdered bodies of her friends along the way and protecting two children. Laurie does not, however, end the threat of Michael Myers (well, no one does for the next ten movies or so). Dr. Loomis, Michael's psychiatrist, delivers the coup de grace to Michael, sending him out the window and ending the terror for this movie, at least. Laurie is, to a T, the model for Clover's definition of a Final Girl[21].

What I find most interesting about Laurie, however, is that although her survival has been interpreted as being due to her virginal state, John Carpenter and Debra Hill have gone on the record indicating that is not why she survived. They have explained that Laurie survived while her sexual peers did not because she was more aware of her surroundings. The fact that she was a virgin was a coincidence. It is amazing how a quirk of interpretation has set in motion a trope that has penetrated the horror genre for decades, influencing the treatment of women as well as the treatment of the genre as a whole. This is an issue that I feel comes up a lot with the horror genre: people go into a film expecting to find something, and because they are looking for it they find it where it does not exist or they make tenuous connections that are opposite to what was intended.

Laurie is not the first Final Girl, though. Sally Hardesty of *The Texas Chainsaw Massacre* "fame" is better situated to take that title. I'm not sure why Sally gets forgotten as the first Final Girl, or as a Final Girl of note. She fits, for the most part, with the identifiers that Carol Clover mentions in her definition. Sally is dressed more conservatively of the two women and her relationship with Jerry does not have the same sexual connotations that Pam and Kirk's relationship does. She cares for the absolutely insufferable Franklin, putting his needs above her own even when it is more practical not to

---

[21] Except she does smoke marijuana and drugs are a Final Girl no-no.

accommodate him. She suffers a great deal in losing her friends, watching Franklin get hacked up, and then in being tortured physically and mentally by four very disturbed men. Sally manages to escape and flag down not one but two cars to rescue her (for some reason the first driver disappears off screen and Leatherface never pursues him) and she escapes laughing manically into the sunrise.

While Sally checks many of the Final Girl attributes according to Clover, she does not do it happily. Although she cares for her handicapped brother, she is not happy about it and tries to avoid him as much as possible. She does not quietly acquiesce to the situation as a "good girl" would and the hostility between her and Franklin is palpably uncomfortable.

One of the differences between Laurie and Sally is that for the rest of the movie after Franklin dies, Sally spends her time screaming and running away while flailing her arms. She doesn't do much of anything to attack the men or defend herself beyond the single attempt at the gas station where she has the knife whacked out of her hand by a broomstick. At first glance, she doesn't contain that strong, heroic spirit we all want to see in ourselves and believe we would act out in horrific situations. Sally's weakness, her lack of unflappability, is what makes her seem like less of a Final Girl. She's not "strong enough" to carry that mantle despite her reactions being realistic and appropriate. I don't know for sure how I would react in the face of a seven-foot-tall chainsaw wielding man wearing a skin mask but having the wherewithal to keep escaping, even while hysterically screaming, is probably just as likely as fighting back or freezing.

Fight, flight, or freeze are all autonomic nervous system responses to threats, which we don't have much control over. Through training and practicing in adverse conditions people in law enforcement, military, and medical professions can try to work through undesired responses like flight or freeze. However, even they sometimes react in ways that are unexpected, like running away from gun fire or freezing at the sight of massive trauma wounds. If people who are professionally trained to handle circumstances like this cannot guarantee how they will respond one hundred percent of the time, why do we expect that regular people with no training would react any differently?

Viewers are probably not as far away from a Sally Hardesty response to danger as they would like to think. Although she wasn't stoic, Sally managed to keep her wits about her just enough to keep fighting and escape. Few people could probably say with certainty that they could do the same. She may not have been a cinematic badass, but Sally held her own with other strengths.

This is an indication that the definition of the Final Girl trope needs updating. It has already undergone some changes with the 1996 meta-slasher *Scream* where Randy Meeks details all the rules of how to stay alive in a horror movie, which includes don't have sex. That aspect of the virginal Final Girl trope is undermined by Final Girl Sidney Prescott, giving horror the opportunity to move past aspects of itself that hadn't aged well over the past decade. The women have a greater variety of names and characteristics. They're not all likeable or good in the typical conservative sense. In some cases, though not a Final Girl, there are female characters who fight

63

back despite being smaller, weaker, or "too feminine" to defend themselves.

Another notable mention who would claim the first Final Girl status from Laurie Strode is Jess Bradford in 1974's *Black Christmas*.[22] Though she is from the same time period as Laurie in *Halloween*, Jess Bradford in *Black Christmas* could not be more different from Laurie. Jess is sexually active and pregnant. She tells her boyfriend, Peter, that she is going to get an abortion and even though he doesn't want her to get one, she doesn't capitulate to his demands. She makes the mistake of thinking Peter is the murderer and she kills him. Jess does not allow things to happen to her passively. She seeks out police assistance and when that does not work, she finds other people who can help her to get the police to take her seriously. Shockingly, in this movie the virginial good girl, Clare, is the first one to die.

There are other early subversions of the Final Girl trope in classic slasher horror movies which were created due to the success of *Halloween*. One of these is Sean Cunningham's 1980 *Friday the 13th* (ch-ch-ch ah-ah-ah![23]). Annie Phillips is the character viewers would have initially thought would be the Final Girl: appropriately dressed, polite, excited to start her job at Crystal Lake. Yet she barely makes it through the first twenty minutes of the movie. Who amongst the remaining three women – Marcie, Alice, or Brenda – is

---

[22] Some articles suggested that Diane Adams from *Silent Night, Bloody Night* (1972) could also be an iteration of a Final Girl, but I would strongly disagree. It seems like those people are using the term too loosely.
[23] According to composer Henry Manfredini, that's not actually the sound in the film. It's supposed to be "ki-ki-ki, ma-ma-ma" which is meant to represent Pamela Voorhees hearing Jason saying, "Kill her Mommy! Kill her!"

supposed to be our Final Girl? All three of them drink, smoke, and engage in copious amounts of exposed skin bathing suit shenanigans with (gasp!) the male counselors. I would not interpret any of them as being particularly virginial. None of them have masculine names or act more like a tomboy than another. Marcie has sex while Brenda and Alice participate in strip Monopoly.

Alice turns out to be the "lucky one" as she stumbles across her friends' bodies and runs away into the waiting arms of Mrs. Voorhees, the real killer. She fights back, trading blows with Jason's mother in comical fight scenes and struggles, until she ends up decapitating Pamela and is adrift on a kayak on the lake. The movie gets in one last scare[24] when the killer's son, Jason, leaps out of the lake to drag Alice under. Thanks to her plot armor as a Final Girl (or that final sequence being a dream), Alice survives only to be disbelieved about the dead boy in the lake trying to drown her. *Friday the 13th* is another clear example where the Final Girl standards are not upheld: first in the killing of the ideal Final Girl Annie Phillips, and then in the survival of a girl who participated in all of the sure-to-cause-death activities as dictated by Clover.

So, is it time to update the characteristics of the Final Girl? One survey respondent would seem to think so, arguing that the likes of Nancy from *A Nightmare on Elm Street* make a far better Final Girl than Laurie (Clover's ideal) due to Nancy actually fighting back and taking agency over her situation, rather than just defending against her attacker. The respondent noted that once Nancy figured out what was going on with Freddy, she took action to keep herself

---

[24] As per the rules from Randy Meeks in *Scream*.

and others alive, and when being on the defensive was no longer an option she went on the offensive. It was Nancy who took down Freddy in the end. She maintained as much control as she could with what she had. Whereas Laurie was constantly on the defensive, running away from Michael, somehow losing the keys to Tommy's house, hiding in spots that are obviously bad, and ultimately having Michael dispatched by "the doctor guy". The respondent goes on to acknowledge that the term Final Girl may have meant something different when *Halloween* came out, and that is why Laurie is held up as the banner girl for that definition. However, the respondent also indicates that for her, the definition of a Final Girl should also be inclusive of someone who has "pizzazz", "takes no shit", and has independence. She's looking for her Final Girls to emulate Nancy, not Laurie.

I'm in agreement with this respondent that the definition of the Final Girl should be, and has been, changing. Carol Clover, however, lamented the new version of the Final Girl who has become "detached from her low-budget origins and messier meanings" and turned into a cleaner and more upscale 'female avenger' " (Clover, p. x) in her new 2015 preface to *Men, Women, and Chainsaws*. She takes issue with this new idea of the Final Girl as a feminist icon or as a hero who survives insurmountable horrors in these movies. Instead, she would prefer them to be referred to as "tortured survivors", "accidental survivors", or "victim-heroes", placing emphasis on the *victim* half of that word. She does not like that these women who suffer and then survive only with a heaping helping of luck[25] are being thought of as people to emulate. She wants us to recognize that although these women might be heroes, they are more victims because of everything they have been through and there is nothing heroic about suffering. Traditionally, according to Clover, the Final Girl has not been a conquering survivor. She's been more like the victim who managed to get away, and this seems to be the way Clover would prefer her Final Girls to remain.

What bothers me most about this position is her suggesting that for new Final Girls, their entire existence only becomes important due to her actions in the final half of the movie when she takes on her role-reversal, becomes more masculine, and takes down the killer. I'm not sure what horror movies Clover was watching up until 2015 when she wrote this preface, but I have found newer Final Girls to have a lot more character and *chutzpah* than previous Final Girls. They also don't typically abide by the "rules" famously outlined in *Scream* (1996) that make a Final Girl "worthy" of survival. Sally and Laurie become far more interesting once they start fighting back and engaging with the killer. *The Descent* (2005), *You're Next* (2011), and the new *Hellraiser* (2022) all have well developed, multi-dimensional characters who break out of the tropey good girl mold (which they aren't even cast in to begin with) of the typical Final Girl and display flaws, fears, and desires before they are forced into dire circumstances.

The Descent's Sarah Carter deals with betrayal from her best friend and her husband, the loss of said husband and her child, and the rekindling of friendships with a group of only slightly less well-

---

[25] From *Men, Women, and Chainsaws* preface "...given the element of last-minute luck (she happens, in her flailing, on a cup of hot coffee or some other such item which she throws into her assailant's eyes...")

developed female characters (who are easily distinguishable from each other). She becomes a feral woman, much like one of the cave dwelling monsters, and manages to escape after exacting revenge against Juno, her best friend who had the affair with Sarah's husband. Sarah suffered much more before the horror started and named her the Final Girl. She and Juno were well defined outside of the terror of the caves. While the other four women are not as fully fleshed out, they are distinguishable from one another through their relationships to each other and their personalities.

Erin Harson in *You're Next* (2011) seems like she might initially be the prototypical good girl. In the car ride to Crispian's[26] parent's house, she is sweet and excited. Then she suggests they pick up alcohol, a Final Girl no-no, and it's clear she harbors some sort of secret when they talk about Crispian's parents. It is revealed after Erin bludgeon's a man to death that she was the daughter of a survivalist father who taught her a few things. Even before this scene, her comment about pouring gasoline into the basement and tossing a lit match down to burn up the occupants lets the viewers know that something is up with Erin. While the other people are screaming and panicking, Erin remains calm and gives orders to try and keep everyone safe. What is notable here is that the other three women die, with the goody-two shoes favorite Aimee dying first. Crispian laments that had Erin been "normal" everything would have gone according to plan. But Erin was not normal, and she also has a good backstory for her strengths which are not infallible. Erin does

---

[26] Gosh, what a terrible name.

not change to become a strong Final Girl; this is just what she always could be but didn't need until the time came.

While Riley McKendry in 2022's *Hellraiser* is exceedingly unlikable with her drug addicted whining and tantrums, this again makes her different from the Final Girls of the seventies and eighties. She has (major) flaws in the form of being a recovering drug addict, having rocky relationships with everyone in her life, and the first time we see her, she and her boyfriend are having very aggressive sex which is probably the biggest no-no of all Final Girl requirements. Despite Riley being unlikable and annoying (this is very much my opinion, I'm sure others like her, please don't @ me), you feel compelled to root for her in the end. She is trying to save her brother who got sacrificed to the Cenobites only because he was trying to help Riley. Riley loses several friends to the puzzle box along the way, all while being blind to her manipulative boyfriend. She is not defined by her final moments in the film, and she may not even really be redeemed by her actions. Those last minutes of *Hellraiser* help to show her as a fully formed character and it is up to us, the audience, to decide if she is worthy of our recognition.

Contrast these women with the Final Girls Clover would prefer. Sally Hardesty is not around for much of the first hour of the movie. Once the group exits the van, we follow the antics of Kirk, Pam, and Jerry as they get picked off one by one. It is not until the last half hour of the movie that Sally starts to show some spunk when she fights with Franklin and then deals with the marauding Leatherface and his cannibal family. Of the two women in the film, Sally is the most modestly dressed in her long white bell bottoms and blue halter top, taking on the role of caretaker for her wheelchair bound brother.

Laurie Strode is largely forgettable in *Halloween* (1974) except for her role as the Final Girl. She is the good girl who studies and has no dates to the prom, but she notices Michael Myers weirdly wandering around Haddonfield. She fights back against Michael when she lucks into finding a weapon like knitting needles or a coat hanger. She does not take out Michael, or even successfully escape without help from Dr. Loomis, a man. She reacts to her environment and the threats within it; she does not act upon it.

The idea of a masculinized, strong, and unflappable Final Girl can also be an inaccurate depiction of, and damaging to, other kinds of Final Girls. Take for instance Barbara in *Night of the Living Dead* (1968) and Wendy Torrance in *The Shining* (1980). Neither would immediately come to mind when thinking of Final Girls. They are maligned, denigrated, laughed at. Stephen King has made it clear that he hates Shelley Duvall's performance in the movie, calling her a screaming dishrag. Ellen Ripley and Sidney Prescott these women are not. Yet they bring something else to the table that is overlooked: realism. Hearkening back to earlier comments about the odds of people actually being able to stand up in the face of sheer terror and survive without a little screaming, Barbara and Wendy react how many people probably would in these situations: they're scared out of their fucking minds and they show it.

But despite their trauma responses, like Barbara trembling on the couch in the house for a portion of the movie, or Wendy being unable to leave her abusive husband and stand up for herself, they still manage to do brave, Final Girl-esque things. Barbara survives the initial zombie onslaught (which her brother does not) and escapes. She aids other characters. She acts upon her environment and doesn't just allow it to act upon her. She may not be the strongest in the group, but her survival instincts are top notch. Wendy, despite her frantic screaming, manages to outwit Jack and a hotel full of malevolent spirits who try to stymie her at every turn. Wendy fights back multiple times successfully with weapons against Jack. She saves herself and her son. Not only is she not rescued by a man, but the rescuing man (Dick Halloran) dies. For all of Wendy's alleged short comings as a Final Girl and as a film character, she acts pretty impressively.

Wendy and Barbara's characters suffer from this idea of the Final Girl who must be strong and calm under pressure and never falter. Although they fight back, they struggle in their fight, which makes people perceive them as weak. Even Carol Clover's version of a Final Girl relies on her ability to act in situations with determination and not with a lot of outward fear. People expect to see a strong woman at the end of film, not someone who panicked shrilly throughout the entire movie. Strength can come in different forms though, and it doesn't always look like the stereotypical strong character who always has an answer or who you know will be able to get out of any bad situation. For much of *The Shining,* I thought Wendy would bite the dust and Danny would be saved by Halloran or survive on his own after Halloran died. Wendy survived and saved her son in the process, managing to beat the odds just like other Final Girls. For all of their supposed shortcomings, the realistic portrayals of trauma and struggle by Wendy and Barbara deserve Final Girl status.

Masculinizing women to be "appropriate" Final Girls is supposedly a way for men to feel comfortable in identifying with them. They have male names, they act like male characters, and they don't have sex (because men don't want to think about having sex with other men). Aside from the fact that I've shown that male name and masculine trait points are not generally accurate, this hurts female characters in film the same way disparaging Wendy and Barbara hurts female characters in film. One specific type of person should not be the only one to survive every time. There are many examples of women who fought back in horror films who could have survived but were too girly or slutty or not masculine enough. There is nothing wrong with having your typical stereotypes in horror films (jock, nerd, popular girl, etc) but it is probably about time that someone other than the tough Final Girl stereotype gets to consistently live until the end.

Evolution is an important part of anything maintaining its relevance in society. If horror movies did not develop beyond gritty, muted slashers with minimal gore they probably would not be as popular as they are today with the wide range of subgenres. (A subgenre for every taste!) That's not to say there is no place for gritty, muted slashers: they continue to be fan favorites, influence new movie makers, and provide the bedrock from which to move forward. Laurie Strode and Sally Hardesty are important for their roles as the initial Final Girls, the ones who paved the way for the likes of Sarah Carter and Riley McKendry. They are remembered and considered iconic for a reason. But a new century calls for a new kind of Final Girl: one who can fend for herself with her flaws, and who also works to defend other characters in the movie.

72

These newer Final Girls, like the original ones, do not allow themselves to become victims of circumstance. They do, however, take more agency in dispatching the villain. They lie in wait for him, rather than the other way around. They may have started out as being the hunted, but they become the hunter. These newer Final Girls also show that there is no "right" way to be a Final Girl. They are human: they have sex, drink, do drugs, make mistakes, and sometimes get people killed. But survive they do, and in spectacular fashion by their own means.

What I have found to be interesting in the traditional research and theories on the Final Girl, is that the Final Girl is perceived as a subversive trope. Men who watch horror films are forced to identify with a woman lead, something not often required in other movies, even today (in the year 2023, to be specific). Frequently, these women take on roles normally held by male characters in other movie genres. They are leaders like Ripley in the *Alien* franchise; headstrong and ambitious like archeologist Scarlett in *As Above, So Below* (2014); the protector, as Laurie Strode is in many of the *Halloween* sequels where she is an adult. These movies put women's issues and concerns at the forefront of the film, even if they are being viewed through the lens of a male director.

The point is that these issues are still being brought up, and even more so when a project is helmed by a woman. In *Hereditary,* issues with motherhood are explored and in *Midsommar* Dani has to reckon with an emotionally abusive and neglectful boyfriend who is ultimately set alight in a giant bear pelt[27]. Even films that would

73

seem to play to the male gaze will often focus on women's issues. *Jennifer's Body* (2009) doesn't sexualize her in the traditional way that women's bodies are sexualized in film, refusing to fragment her body with the lens. Jennifer takes out a very real and reasonable female rage on men who would only use her for her body. She owns her sexuality and uses it to her advantage while refusing to occupy a smaller and quieter space as is socially expected of women. Newer horror tries to remove the judgement and stigma typically aimed at women who are viewed as emotional, hysterical, or who partake in certain activities (drinking, drugs, sex) or jobs (pornography like in *Cam* (2018)). Small steps like this can be influential towards changing sexist or judgmental opinions towards women in general.

Though these women may be masculinized via traditional beliefs of what is masculine or feminine, having the ability to survive is not a masculine trait. In the time period when the Final Girl trope first appeared, gender expectations were a lot different as compared to now. It could very well have been subversive to have a female lead the same way that *The Twilight Zone* was subversive for the themes it had in its episodes. Now, though, expectations of men and women have changed and what once was subversive no longer has that role. Times are changing and so are the movies we watch and the roles people, especially women, play in them.

The importance of the Final Girl lies not just within her value as a trope, as a way to direct a story, or as the standard bearer

---

27 Don't ask because I don't know. I didn't even like this movie, and didn't really "get it" but I did like when the annoying boyfriend died and Dani was finally happy.

for horror. The way the Final Girl, and women in general are represented on the screen matters in the greater context of culture and life in general. When girls and women are seen on screen being brave, doing hard things, or doing certain jobs it encourages women in real life to be those things. Research done by the Geena Davis Institute and the J. Walter Thompson Company (2022) have found that 61% of women globally said that female role models in film and television were influential in their lives. 90% of women feel that female role models in media are important. A particularly startling statistic is that one in nine women globally found the courage to leave an abusive relationship based on positive female role models (Geena Davis Institute on Gender in Media, 2022).

Media being influential on the lives of girls and women is also seen in the Abby Sciuto Effect. Abby worked within NCIS as a forensic scientist, doing what used to traditionally be a man's job. Since she has been on screen, more girls have sought out careers in forensic science. Abby showed that math and science, typically male coded job options, were also viable for women (CBS, 2018). This can also be seen with Dana Scully (who also has the "Scully Effect"), Temperance "Bones" Brennan, and even cartoon or fantasy characters (Geena Davis Institute on Gender in Media, 2022). In 2012 after *The Hunger Games* and *Brave* came out in theaters, young girls flocked to the sport of archery, citing Katniss and Merida as their main influences for picking up the sport (Heldman, 2016).

What we consume through popular culture and media is incredibly important. Updating the Final Girl to be more modern and to have more agency can potentially have a huge effect on the way girls and women perceive themselves and their own abilities.

As I mentioned way back in the Introduction, Ellen Ripley was a major role model for me. I wanted to be brave like her and solve problems like her. Imagine the untapped potential that could be accessed if more young girls could feel that capable.[28]

My favorite response to the survey question of is a Final Girl a survivor or a victim is one that reflects the idea that we should be updating the definition of what a Final Girl is: "Both, the final girl is an archetype of losing a piece of femininity that is rediscovered through adversity." The Final Girl is a woman triumphing through adversity, losing parts of herself and coming back as something both stronger and irrevocably changed.

---

[28] I couldn't find a way to fit in another article that I read, but check out the citation for an article written by Majdoulin Almwaka.

The Tangents

Do Final Girl characteristics mimic the "Ideal Victim" characteristics looked for in law enforcement?

"Blondes make the best victims. They're like virgin snow that shows up the bloody footprints."
-Alfred Hitchcock

There is a concept known in law enforcement and victim advocacy as the "Ideal Victim" which is when victims must fit certain criteria in order to be taken seriously when coming forward about a crime. It is a cultural and social construction that identifies which people are most worthy of receiving help for being victims or who can be viewed as a victim. It is a weird thing to look at people and decide who would be a "good" victim and you might find yourself balking at the idea. But think about the last time you saw someone on the news who claimed to be victim of a crime and you did not second guess their story or whether the alleged victim may have played a role in their victimization.

This most frequently comes up with women who accuse others of rape or sexual assault. Can you think of times when, maybe not you specifically, but others have tried to neutralize the victim's accusations by saying they had too much to drink, or wore the wrong clothes, or were somewhere they did not belong? What about if the victim of a crime was a prostitute, or drug user, or a man? Would any of those things have you second guessing their story?[29]

There are five basic characteristics of an ideal victim: 1) the victim can be perceived as weak or sympathetic (elderly, young, female); 2) the victim is doing something respectable or legal at the time of their victimization; 3) the victim could not be blamed for their victimization (not drunk or high, not out after dark); 4) the offender is unambiguously bad (career criminal, not an upstanding college student); 5) the victim and offender were strangers.

These criteria could be applied to horror's Final Girls. Final Girls, historically at least and sometimes still, are supposed to be virginal good girls. They engender sympathy because bad things

---

[29] I tried to be as general as I could and will be as general as I can when speaking of real-life victims of crimes because it is not only women who are crime victims, even in cases of sexual assault and rape. Although men commit over 70% of all crimes, they have also historically been victimized at higher rates for violent crimes. More recent statistics suggest that this is changing and that victimization rates for women are rising in certain crime areas. I'd also like to note that although rates of rape, sexual assault, and domestic violence are still consistently higher for woman than for men, male victims of these crimes **do exist** and part of the issue is the lack of reporting on the part of men due to stigmatization and their lack of being an **ideal victim** for these crimes.

shouldn't happen to good people. One would not expect to find them wandering the streets late at night cavorting with boys and doing drugs. The villain in horror movies is almost always unambiguously bad. Even when there is a sad back story to sort of explain why he might be out murdering people under the cover of darkness, it usually is not enough of an explanation for the audience. Final Girls would make the perfect versions of the traditional ideal victim.

On the final point, the people being murdered usually don't directly know the person or people murdering them. 2008's *The Strangers* is about as perfectly anonymous as you can get. Even the poorly done 2018 sequel, *The Strangers: Prey at Night*, provides a perfect non motive when Kinsey asks why the strangers are doing this and gets the response "Why not?" How could anyone not feel sympathy for the characters in the face of such directionless violence?

In some cases, the victims know of the villains through urban legends or folklore, such as in *A Nightmare on Elm Street* (1984), *Candyman* (1992), or *Urban Legend* (1998), but they're still not directly linked. No one blames the Final Girls or boys for killing the big bads in these movies. While Candyman in the 1992 film, played by the amazing Tony Todd, had one of the more sympathetic back stories, the viewer would still have a hard time feeling like he should continue to exist and terrify the Cabrini-Green Projects.

In more recent movies, some of these ideal victim rules get subverted or completely ignored. The movie most famous for subverting many of the horror genre's tropes is *Scream* (1996). The Final Girl has sex and drinks and still survives. Their friends are the ones doing the killing. This crosses out criteria two, three, five, and maybe four.

The movies *You're Next* (2011) and *Ready or Not* (2019) also cross off some of the criteria. Erin reveals that she grew up in a survivalist camp in Africa about halfway through *You're Next*, removing the criteria that she could be perceived as weak, and also potentially her ability to be seen as respectable depending on how someone views the survivalist community. She also knows everyone who is trying to kill her. *Ready or Not* has Grace Le Domas fighting back against another eccentric family, making her a strong character but also maybe a more sympathetic one as a former foster child. She again knows the people who are trying to kill her: her new husband's family who have a pretty interesting motive.

One of the major drawbacks of a culture that embraces the philosophy of an ideal victim and a pure Final Girl, is that if you do everything right you should be able to avoid any sort of pain, suffering, death, or shame. When a woman is identified as a Final Girl, that means that she is supposed do everything right and escape the villain. When someone is an ideal victim, it means they are more deserving of assistance and compassion because they didn't do anything to deserve what happened to them. Realistically, however, bad things can and do happen to people who do everything right and victims of crimes are not always "ideal" but are still deserving of belief and compassion.

In *Wolf Creek* (2005), Liz and Kristy arguably do everything right and by the book for being a Final Girl. They are dressed conservatively, are not traveling with boyfriends or engaging in sex, and are driving around doing very acceptable tourist things. Once captured by the equally terrifying and comedic Mick Taylor, they make smart decisions [30], and they seem poised to escape, at a minimum, and at most potentially get the police involved. But that doesn't happen. Neither girl survives, no matter how much you yell at the movie or rail about how they fit the Final Girl aesthetic. Instead, Ben, who we have largely forgotten about and written off as dead because he is not the Final Girl and we really don't care about him, is the one who gets away. But the truth of the matter, which comes clearly across in *Wolf Creek* and other movies, is that real life does not care about the Final Girl or the ideal victim. It is chaotic, unpredictable, and sometimes even when you do everything right, you still die.

The concept of the ideal victim has begun to fall by the wayside but can still be seen in the criminal justice system. The changing standards have a lot to do with the changing cultural standards of society. These changing cultural standards can also be seen in the topics that horror movies choose to tackle and in how characters are shown and treated. Final Girls from seventies and eighties horror movies mostly reflected the ideal victim criteria above. Final Girls now look a little different than their predecessors, having evolved along with the standards of what it means to be a good girl, and sometimes showing how irrelevant those standards really are.

---

[30] Except for Liz shooting Mick only one time. She should have at least done a double tap, smashed his head in, anything. I was screaming at the movie for her to ensure his death and this frustrated me so much.

When You Gaze at the Man the Man Gazes Back at You...or Something[31]

**"I've seen enough horror movies to know that any
weirdo wearing a mask is never friendly."**
-Elizabeth, *Friday the 13th Pat VI: Jason Lives*

The term "male gaze" hadn't been something I was aware existed until a friend of mine mentioned it after watching the 2017 *Wonder Woman* movie directed by Patty Jenkins. Her comment was along the lines of the *lack* of the male gaze because Wonder Woman was not sexualized at all in the film (despite her revealing armor) and there is one point where she does the "superhero landing" (thanks for that one Deadpool) and her legs jiggle. My friend was most excited about that because, according to her, you never see a woman's body parts jiggle in movies unless it's her breasts jiggling. Jiggling in

---

[31] "Whoever fights monsters should see to it that in the process he does not become a monster. And when you look long into an abyss, the abyss also looks into you." *Beyond Good and Evil*, Pt. IV, 146, Frederich Nietzsche

other parts of the body means that the woman is fat, which is unattractive to men and the "male gaze".

The definition of the male gaze from Oxford Reference is "a manner of treating women's bodies as objects to be surveyed...see also objectification". As defined in film theory: "the point of view of a male spectator reproduced in cinematography and narrative conventions of cinema, in which men are both the subject of the gaze and the ones who shape the action and women are the object of the gaze and the ones who are shaped by the action." The word choices are important here as the man is the *subject* of the gaze and the one who *shapes* the action, while the woman is an *object* and the one *shaped* by action. Men are the ones with agency in films, and women just happen to be there to react to whatever the men are doing.[32]

At this point you might be wondering what this has to do with horror movies and when I'm going to get on with it already. Misogyny in film is all about the male gaze and how men present and interpret information about women in film. So, please bear with me as we travel through a quick history of "the gaze".

The idea of a gaze is not that new, having been introduced by Jean-Paul Sartre in 1943 as the idea that gazing at another turns them into an object and creates a power difference between the person gazing and the person being gazed at. Laura Mulvey takes this idea further when she defines the "male gaze" as men looking at

---

[32] There is something called "The Sexy Lamp Test" where if you can replace a woman with a sexy lamp and not significantly change the story line, then you don't actually have any women in the movie. If a woman's only job in the movie is to relay a message, then she is a Sexy Lamp with a Post-It Note.

women to achieve some sort of pleasure from them, specifically when women are in passive roles. Most films are directed, produced, and written by men, which creates media formed entirely by the desires and perceptions of men. Alfred Hitchcock is particularly famous for his use of the male gaze in his films, training the camera lens on the women as if they are being pursued.

Take, for example, *Jennifer's Body* (2009) a film written and directed by women, and in which women are the primary characters. The film is ultimately supposed to be about female friendship and a commentary on how society views women: a feminist film for women. Yet the trailers for the movie are geared directly towards young men, showing a kiss between Amanda Seyfreid's Needy and Megan Fox's Jennifer. Jennifer's body and sexuality are front and center in these trailers, making her the object of the gaze, something to passively please men. In reality, Jennifer is using that male gaze to entrap, kill, and eat men who are only interested in her for her looks. The movie bombed because it wasn't what people (men) expected, but it later developed a decidedly feminist cult following. It was probably disconcerting for young men to go to a movie they thought would be filled with girl-on-girl action only to see the teenage boys torn to pieces.

One of the other things to note in *Jennifer's Body* is that there is very little fragmentation of Jennifer: she dominates scenes not with her legs or breasts or face, but with her entire character. While many films rely on fragmenting a woman – showing only bits and pieces of her at a time, especially during scenes where she is being murdered or engaged in sexual activity – Jennifer does not suffer that same treatment. She's a whole person, not just her titillating body parts. This is important because men are rarely fragmented, even when they are killed or having sex. They take up space on the screen.

A movie which immediately comes to mind that fragments a woman is Eli Roth's *The Green Inferno* (2013). The Final Girl, Justine, is about to be subjected to female genital mutilation and while her face is shown repeatedly and almost constantly in the lead up to the mutilation, showcasing her terror and panic, when the actual act is about to happen and she starts to struggle, the camera begins to focus primarily on her body. She is broken down to her legs tied to posts and her breasts as her chest heaves, her flat stomach above the burlap undergarments she is wearing. Justine becomes only her body in the throes of panic and the very human fear in her face is removed. Fragmentation serves to dehumanize, making it easier to see a person as a collection of body parts with no feelings attached. Dehumanization makes it easier to inflict harm on a person or watch harm being done to them.

This sequence combines sex – Justine's writhing body – with fear. Why, though, would I consider this sexual? I would answer this with another question: Why is it necessary for Justine's breasts to be so prominently featured when they have nothing to do with what is happening?

Both sex and fear have prominent roles in the horror genre. The sex seeking part of the male gaze dictates that women be decorated in tight, skimpy outfits with heels and perfect make-up, and the fear seeking aspect of the male gaze demands that they be terrified. You don't really think it's coincidence that when women are getting chopped up and are being tortured in horror movies that the camera fragments their bodies or focuses heavily on the curves of the hips and breasts, do you? Think of the men that are shown on screen before, during, and after they are killed. Are they half naked with their muscles featuring prominently in shots where they are running away or straining against bonds as a killer slices into them? Does the camera linger on their faces or their bodies? Are they shown as a whole or in fragments?

Back to *The Green Inferno*[33]. Final Girl Justine has most of her clothing removed repeatedly throughout the movie. Samantha attempts to escape and is literally fragmented into pieces of her tattooed skin which the children wear like temporary tattoos. All three of the remaining women are made to be remove their pants and are violated by a virginity test in front of the entire village with the camera's focus remaining primarily below the waistline. The men, meanwhile, remain clothed the entire time. Even when Daniel is strung up on a pole where all of his bones are broken and he is then smeared with some green substance for ants to chew on him, he keeps all his clothing on even though it would definitely have been easier for the ants to do their job without Daniel wearing clothes. Why not strip Daniel naked?

In the *House of Wax* (2005) remake, both women remove their clothes and Paris Hilton, as Paige, has an extended chase scene in her underwear and a robe which has strategically fallen off one of her shoulders. When she dies, the killer uses a camera to focus in on parts of her body, finally settling on the pole that has impaled her

---

[33] Don't get the wrong impression. This is actually a pretty good movie. It's also far better than *Cannibal Holocaust*, from which Roth drew the inspiration for his film, and which I would never recommend to anyone ever.

face. Her boyfriend, who moments before was about to have sex with her, dies fully clothed with no lingering shots on any specific part of his body. The other four men that die (the two antagonists Bo and Vincent, and the friends Wade and Dalton), have no lingering fragmented shots of their bodies. Bo and Vincent are repeatedly shown "conjoined" in their death. Dalton has his head neatly chopped off after a short chase scene. Although we do sort of vaguely see Wade naked as he is getting coated in wax, his body is shrouded in shadow and his death comes later as his body sinks into the melting wax floor. Dalton's death is probably the most stylistically inclined out of all of them but doesn't match up to the length of time we spend with Paige.

An example that most bothers me is with Ripley in *Alien*. She has defeated the alien menace (at least, we think so at this point) and rescued herself and Jones the cat, setting off in the escape pod. She strips down to get into her space sleep outfit which consists of a white t-shirt and white underwear. I don't know why she couldn't have white shorts like the men, but fine. That's not the most annoying part of this movie which has, for the most part, escaped the male gaze and allowed Ripley to be a fully formed character with her own badass agency separate from her being a woman. It is at this point that Ripley leans over equipment and her butt crack is there on display.

For what purpose? It is clearly noticeable, and I would not accept that it was an "innocent" accident. If the underwear were too small, why not get a larger pair? It is almost as if Ridley Scott had to somehow compensate for the fact that out of everyone who survived, it was a woman who had not had to get naked or appear weak throughout the film. None of the men in the film are shown in such a compromising or potentially sexual situation, so why Ripley?[34] Why now, in the very end of the film? The fact that she is vulnerable, in her underwear, getting ready for sleep, was clearly conveyed without the need for her butt to show.

Something that came up again and again in the research was that violence perpetrated against women in the service of the male gaze was a way to punish women and satisfy male viewers, presenting them as sadists[35]. The movies allegedly blur the lines between male pleasure at seeing a naked or skimpily clad female form, and the pain she is suffering at the hands of a villain who is usually male. What does this mean for male viewers of horror? Do they truly love seeing women suffer? What about female viewers? Why would they revel in the pain of other women?

Women take twice as long to die on screen as men do (Molitor, F., & Sapolsky, B. S., 1993). Sometimes this is due to an extended chase scene where the woman screams, falls, gets up, fights back, but ultimately meets her demise as Paige does in the *House of Wax* example above. Once the woman is dead the camera will linger over her dead form, observing her like a piece of art instead of a murder victim. Does this suggest that the viewer is sadistic and revels in the torture of women? What about in films like *Hostel*

---

[34] I did read an article that said Ridley Scott suggested that Ripley was sexually assaulted by the robot, Ash, when he tries to shove a rolled-up magazine down her throat in the scene where he goes crazy. I haven't found any direct sources of Scott making this statement, and it also seems to go against other aspects of the movie, such as having a man be violated by the Facehugger. I always looked at this scene as Ash malfunctioning and just grabbing whatever was around and acting crazy with it.
[35] Some believe that the male gaze can *only* be sadistic.

(2005), *Saw* (2004), and *The Green Inferno*? Yes, I know, I just roasted *The Green Inferno* for its fragmentation of women and excessive focus on female genital mutilation. But, for all of that, the most gruesome death in the entire movie is Jonah's at the cannibal camp. After having his eyes popped out (AND EATEN), then his tongue sawed out of his mouth, he is dismembered while alive, flopping and screaming on the sacrificial table. The other men die similarly gruesome deaths, while the women do not. One woman gets a spear thrown through her head; one woman's death we do not see (we only know she is dead because the children are "wearing" her tattoos); and one girl cuts her own throat. Much more tame than what happened to Jonah.

Hostel is the systematic torture of a group of young men traveling around Europe. Some of the people who torture them are women. *Saw* focuses on the psychological and physical destruction of two men while showing other men dying in heinous torture traps. The one woman who gets caught in a trap survives without suffering much physical damage (but the psychological damage will be a doozy).

While these movies are outliers, they are all directed by men. They are putting male actors through heinous situations with no real happy ending for them. How does the male gaze play into these movies if horror movies are supposedly catered exclusively towards men and the male gaze? Is it just that men enjoy watching suffering in general? Do they squirm and get nauseous and uncomfortable watching these scenes? Would they feel the same way about women suffering through the same thing?

One idea may be that we enjoy watching other people suffer in film and television because it is "safe". We know that these are actors and at the end of the day they will go home usually none the worse for the wear. It is like riding a roller coaster: we know we will be scared of the drops and turns and speed, but at the end of the day we will still be able to safely get off the ride. No one would say that people have a death wish for enjoying roller coasters, even though sometimes they think they might die on some of those more intense loops and turns. Horror movies allow us to feel uncomfortable and disturbed, but they always end (even the sequels have to end at some point). So, it stands to reason that we are not sadistic or enjoy watching other people suffer simply because we enjoy being able to safely experience those feelings without ourselves or other real people suffering any consequences.

I think this idea of horror movie watchers enjoying these movies because they enjoy seeing others in pain is largely overblown, even for male viewers. Using horror movies as a safe way to feel fear is a better concept, and one that makes sense even for men. While it is not culturally acceptable to show fear in general, men are expected even more so to not show fear or any vulnerability. Fear is weakness for men, and men are not allowed to be weak. Being able to watch a movie and for a little while be able to identify with a person who is suffering as a way to suffer themselves without other people realizing, could be an important emotional outlet for men.

Does this mean that the male gaze isn't a real thing? No, it is definitely a real thing. The fragmentation of women and the lingering camera shots of their bodies (in all movie genres, not just horror) showcases the male gaze. The concept that I'm suggesting is not an actual thing is the sadistic male gaze viewer who revels in another's pain. The male gaze in horror, specifically with fragmentation, may be an important way for men to identify with the terror of women without having to actually identify with women. While this is also a problematic idea, it is far more preferable to sadistic men. I would suggest that there are many reasons that people enjoy horror movies other than for the suffering.

It cannot be denied that there is likely a subsection of male (and probably female) viewers who get more out of watching women being harmed than men being harmed. After all, how many times have we watched men plead for their lives in movies versus women? Men often die in the protector role, trying to be macho, not screaming for their lives. Is it because society finds it weird for men to be less than masculine (or "more feminine") on screen? This hits home when you read that Dario Argento prefers to watch women with "a good face or figure" be murdered rather than an ugly woman or man. Pascal Laugier (who did one of my least favorite movies, *Martyrs* (2008), and one of my favorite movies, *Incident in a Ghostland* (2018)) claimed that male characters make things too real for him and prevent his escape from realism, whereas female characters fit into the storytelling he wants to do.

Compare this with the thought process of the crew involved with Ridley Scott's *Alien* (1979). The only person to be essentially raped and impregnated by the facehugger is a man. The crew felt that doing this to a woman would be too easy and would not be as impactful as subjecting a man to these violations.[36] Given the horror

92

genre's penchant for torturing the women, it does seem almost a little too easy to use women over and over again as the receivers of torture and abuse.

While Argento's comment is obviously trending towards misogyny, Laugier's comment is a little more murky. What exactly is the storytelling he is attempting to accomplish? In *Martyrs* two women, Lucie and Anna, break into someone's home in order to exact revenge for considerable abuse they faced at the hands of the homeowner. Lucie slaughters the entire family, plagued by a demonic entity which follows her, and which is later revealed to be an imaginary manifestation of her trauma. Lucie kills herself, and Anna discovers that the homeowner was part of a secret society that was attempting to torture people into transcendence so they could experience heaven in the pure way that only martyrs can. How would this movie have been different if the victims had been male? The villain characters in the film explain that women are used exclusively in the attempt to become martyrs because they feel more pain and tolerate pain better than men do, and thus would be able to handle the journey to martyrdom better than men[37]. It seems like a compliment, but also just an excuse to torment women.

Incident in a Ghostland centers around Beth and Vera who are subjected to sexual, physical, and mental abuse at the hands of two home invaders known as Fat Man and the Candy Truck Woman. Beth experiences dissociative trauma responses where she retreats

---

[36] Which is why people interpreting Ash's magazine assault in *Alien* as a sexual assault seems far fetched.

[37] There is a ton of interesting research on this, too much to address here, but the boiled down version is that women do feel pain differently than men.

into a fantasy land in her mind, leaving her sister, Vera, alone with their tormentors. The two girls eventually escape, are recaptured, and after another fight, are saved by state troopers. What the two girls are subjected to is frankly disturbing and is extremely uncomfortable to watch: they are physically and sexually assaulted, made to watch each other suffer, and endure a level of entrapment that torments the viewer as much as the characters. How would two boys being in this situation have made the movie any less impactful or brought more realism into it? If anything, seeing two boys in this position may have made the movie even more horrifying and disturbing as males are not typically subjected to the kind of sexual and psychological damage seen in this film. Laugier misses an excellent opportunity to plumb the overlooked aspects of male terror, weakness, and vulnerability because he somehow feels that women are more appropriate for the torment in his movies.

Argento and Laugier's comments are especially interesting when looking at the impact of the *Hostel* and *Saw* franchises which spawned many sequels and seemed to make a great impact on society even with the first two installments focusing on the torture of men. In the first *Saw* movie, every single character who dies is male. One is killed by a woman (Amanda). In *Hostel*, the majority of the tortured and killed people are also male. Some of them are even tortured and killed by women. While watching these movies, the tension felt by the viewer does not lessen because the characters are male. If anything, having two women locked in that initial room in *Saw* may have been distracting because there might have been more of a focus on their bodies (as horror is wont to do). A female cast in *Hostel* may have resulted in the viewer feeling more sympathy specifically for the women instead of disgust as a whole for the underground torture and kill industry.

Alfred Hitchcock, a horror legend and inspiration for many, commented, "I always believe in following the advice of the playwright Sardou. He said 'Torture the women!' The trouble today is that we don't torture women enough." (Lord, 2020). In what way did he mean? Hitchcock was allegedly notorious for torturing his female actors off screen. Tippi Hedren has made accusations that Hitchcock sexually assaulted her, ruined her career, and made her miserable on and off set. Additional allegations from other actresses have surfaced over the years, and his biographer, Patrick McGilligan, noted that Hitchcock had a penchant for groping women and other questionable behavior (White, 2021). Torture the women indeed. The 1950's were also hardly a time of great freedom, safety, and enjoyment for women.

In looking at films that torture women and are a reflection of the torture women go through in real life as well, *Rosemary's Baby* (1968) shows the struggles a woman goes through trying to maintain and have a family. Rosemary is continually gaslit and controlled by her husband and her (very creepy) new neighbors. Eventually, her gynecologist, the person a woman should be able to trust without hesitation when pregnant, also betrays her. In as vulnerable a state as pregnancy makes her, Rosemary is taken advantage of, lied to, and manipulated again and again by the people meant to be closest to her. Pregnancy in real life can also be scary and confusing, even without a gaslighting husband who drugs you so you can get impregnated by Satan for some weird cult. Women are subjected to being poked and prodded, having invasive procedures done, and losing all sense of dignity throughout pregnancy, childbirth, and after as people judge her for every little thing she does.

Torture the women? I'd say the women are being, and have been, tortured more than enough over the centuries.

An example of a male director who has not participated in the male gaze is Mike Flanagan. The director of such films as *Hush* (2016), *Doctor Sleep* (2019), and *Oculus* (2013) also directed *Gerald's Game* (2017). This is significant for a few reasons. It was considered to be one of Stephen King's most unfilmable books because it takes place primarily in the mind of Jessie, the main character, who is handcuffed to a bed after a failed sex game with her husband, Gerald, where he essentially tries to rape her, suffers a heart attack, and dies.[38] Handcuffed to a bed in a silky negligee, Jessie should be the picture of the male gaze with all of these opportunities to focus on her exposed skin, her legs, her breasts. But Flanagan does not do this. Jessie is fractured only in times of stress when that is the most important body part for getting her out of a bad situation: her hand as she reaches for a glass of water, her foot as she kicks away a hungry dog. Even in those scenes we see Jessie's face contort in terror and frustration as she struggles, screams, and fights to escape. In what could easily have been one of the most sexual movies, Flanagan takes all of the sexiness out of it. And that is amazing.

All is not lost to the male gaze and directors who can't seem to muster up the imagination to cast males into desperate and

---

[38] I thought the whole sex game thing and being handcuffed to a bed was why it was unfilmable, but apparently not. I read the book prior to watching the movie and found it much less enjoyable than the movie.

vulnerable roles. Combatting that gaze with female directors who do not do these things or having female characters with more agency and diverse personalities can help. Being willing to have a greater imagination and putting males in roles to suffer can also even the playing field, while also tackling some interesting new ideas about how we view masculinity. The gaze of the person writing and directing movies will always be evident, the same way as how we see and interpret the world through our individual lenses. But if we can educate others on these things, and how having fully formed characters of all genders can make movies better, maybe we can have fewer extended chase scenes of women with big breasts bouncing through the forest away from a knife wielding murderer.

The Tangents

The Case for Paris Hilton

"Yeah, let's go investigate the strange smell."

Paige, *House of Wax* (2005)

Paris Hilton, who plays Paige in 2005's *House of Wax* is one of my favorite horror movie women. Yes, ditzy, blonde, velvet track suit wearing Paris Hilton is a badass in my book. She definitely is not your typical Final Girl: she's very feminine; she is sexually active based on the brief sub-plot of a pregnancy scare; her and boyfriend Blake almost have on-screen sex leading to her wearing nothing but underwear and a robe.

But Paris' Paige doesn't run screaming, crying, and tripping into the night only to get speared from behind by the antagonist. She kicks off her attacker, runs to a warehouse, and quickly finds herself a weapon which she effectively uses against the antagonist, if only for a brief and triumphant moment. Of course she dies, since she is not the boyishly named and slightly masculine Carly character[39], our

---

[39] Of note, Carly changes out her slightly dirty yellow shirt after nearly falling into a roadkill pit for her brother's white wife-beater. It provides an

designated Final Girl, but she puts up a fight that I'm sure no one expected her to put forth. She would have been right at home as a Final Girl in *Scream* (1996) and although she doesn't make it to the end of the film, she is an interesting character. In a genre where women get entirely nude all the time Paris subverts what probably everyone expected her to do and she is never seen naked. Paige is likeable, supportive of her friends, and smarter than I think viewers would have thought. She fares better against the murderous masked man than some of the other male characters.

In the 2022 Blue-ray release of the movie, it includes a special feature titled "Die My Darling" where Paris talks about her role in the movie. She speaks highly of the role, and how much she enjoyed participating in it despite being scared to watch the film in its entirety when it originally came out. She mentioned how it would have been "cool" if her character had been running away from the killer in red stilettos (to match her lingerie) though the director nixed that as he wanted the movie to be "realistic". As though a movie where people somehow survive being encased in boiling hot wax for hours or days after is so very realistic. I just can't see anyone packing stilettos to go on a camping trip, but this is Paris Hilton we're referencing here.

Arguably, the character of Paige would be right at home with some of the more recent Final Girls, and might even have made it to the end of the movie had the *House of Wax* remake been filmed a

---

awkward moment where she is forced to change in front of the creepy hillbilly character, and also serves as a visual representation of Carly shedding her femininity for the masculinity she will "need" to help her defeat the Sinclair brothers. Why else take off a shirt that was barely dirty?

decade later. Alas it was not, and Paige bit the dust. Maybe the next *House of Wax* remake will be different.[40]

---

[40] Because let's face it. Every horror movie eventually gets another remake.

Gaslight the Women

"I think we'll start with a reign of terror."

***The Invisible Man* (1933)**

One of the more frustrating tropes attached to the Final Girl is that no one ever believes her. Like, never ever. By the time people start to pay attention or give the woman's spook vibes some credence, it's too late. Someone or many someones are dead, the family pet is missing, there is a person inside the house, or the evil spirit has already manifested. Generally, if the woman had been believed from the beginning, none of this bad stuff would have happened, or at least some of it could have been diverted.

The obvious reason for this is that it helps to create the artificial conflict that will move the plot along and provide a story. Fifteen minute horror stories would unlikely be satisfying to today's audiences (I find the sweet spot is about 75 to 90 minutes) unless they were done extremely well. The other more insidious reason is that women are often not believed or seen as credible in real life. Frequently, women's concerns are dismissed with throw away lines like:

"Are you on your period?"

103

"Women are so emotional."

"Women are irrational.

"Stop being so dramatic/stop overreacting."

A woman's anxiety or concern is perceived as unnecessary and unrealistic – an unfortunate by-product of a hormonal brain that leads to irrational thinking or decisions. There is no crazed stalker killing people or evil spirit giving you nightmares. This mirrors the disbelief women face when they talk about sexual assault or harassment in real life.

What is it about gaslighting that is so essential to horror? Is the isolation of the Final Girl supposed to up our own tension and anxiety, to make us more invested in her success? Does it make the movie more interesting to have a one against all (or most) perspective? Or would we rather see a concerted group effort in figuring out what horror waits in the dark, a la *It Follows* (2014) or *Final Destination* (2000)? It's a little bit Scooby-Doo when the characters get together to combat the evil force as one, but in some cases, it works out. The movie takes on a little more of the mystery or thriller aspect once the protagonist is no longer alone in their struggle to survive and make others believe. But which version is more satisfying? There is no payoff if there is no tension, and yelling at the characters that they need to believe the danger is real can be fun. But it can be tiring if not done correctly. At a certain point, we just want the other characters to believe our protagonist and help them.

While the 1944 movie *Gaslight* has a woman being gaslit into believing she's crazy, Roman Polanski's *Rosemary's Baby* (1968) does a terrifyingly excellent job of gaslighting Rosemary into believing everything is perfectly fine with her life. She repeatedly asks for help from people that she trusts, and other than her one friend who is killed by the cult, everyone tells her that her crazy beliefs are because she is pregnant. None of them, because of their own nefarious purposes, want her to understand what is happening. The movie is not just about the cult and Rosemary's pregnancy with the antichrist. It's also about how Rosemary is systematically broken down until it is too late for her to escape, reflecting real life cult indoctrination.

Leigh Whannel's 2020 movie *Invisible Man* emphasizes issues of violence against women and women not being believed, resembling the original source material in name and invisible character only. While he hadn't intended on the movie going in that direction, Whannel felt that was how it naturally evolved through his conversations with women about the subjects of stalking and assault. Cecilia, the gaslit woman throughout the film, is taken advantage of by her ex-husband in life and after his supposed suicide. When she goes to her friend and sister about her concerns that the ex-husband is still alive, they essentially tell her that it is all in her head and believe her to be hysterical. Her sister dies due to her disbelief, and her friend almost gets his house set on fire as a result of his disbelief. As with many women in horror, Cecilia sees the danger that everyone else is blind to and calls her crazy for believing.

Whannel acknowledges that his film, and horror especially, is a great vehicle for showcasing systemic issues in society and highlighting our collective fears. Women are typically not believed when they accuse a man in power of assaulting her – they are trying to get his money, or ruin his reputation, or they are just jealous of his success. Women also face judgment when living in or trying to get out of abusive home situations. The message of this movie in 2020 was extremely relevant considering the 2017 accusations against Harvey Weinstein; Brett Kavanaugh's assignment to the Supreme Court even after accusations of assault were brought up by Christina Blasey Ford in 2018; and Larry Nassar being sentenced to a total of over 200 years in prison in 2017 and 2018 for sexually assaulting Olympic athletes. Although these women were mostly believed (after decades of silence), the #metoo movement and the destigmatization of speaking up against sexual violence are still works in progress. The comments on Krystie Lee Yandoli's article, "How "*The Invisible Man*" Shows the Horror of Not Believing Women in the #metoo Era" makes that all too clear:[41]

> *"With the whole Amber Heard/Johnny Depp thing going on right now I feel like "believe women" might be a tough theme to shove down people's throats...especially when the female protagonist straight murders her husband at the end even after she knows he wasn't the Invisible Man. Justice? How about Just-leave-him?"*

> *"Believe women, lol, yeah, sure. How about no and wait until proof of something, anything, comes in. And yeah, the whole Amber heard is the perfect example of what happens when accusations are enough to convict, no due process. What a joke"*

---

[41] By the time I was ready to enter these comments into my own text, they had been removed from the article. Thankfully, I had printed the article with the comments when I first came across it. I have typed out the specific comments word for word with whatever grammar and spelling it contained here but have scanned in what I originally printed from the article which is shown in the appendices section.

"Yet another political smear campaign of a good movie. Movie critics have lost all credibility nowadays, and should be viewed as nothing more than whining crybabies. Not everything has to be political, just a good movie. Doesn't help that the whole #MeToo thing is just a witch hunt now. Unless there is proof then they shouldn't be believed because of how harmful it is to be wrongly "MeToo" someone."

"I'm not sure who is more stupid; the morons complaining that you gave away the plot after saying there were spoilers or your commentary, which is some of the worst, factually incorrect, woke filled dross I've seen lately, and that's saying something in these days! How the hell you were given a job as a journalist is beyond me..."

"Yes...if your ex learns how to become invisible and mess with you, this movie will be a perfect lesson about believing women who find themselves in that unique fictional scenario...."

Frankly, I'm not even sure if all of these people watched the same movie as the rest of us.

In the end, Cecilia takes her abuser's power and weaponizes it against him, something many women may wish they could do[42]. Although not technically a Final Girl as Cecilia is existing in a psychological horror rather than a slasher horror, she does manage to triumph over the villain in true Final Girl form. Cecilia is also the focus of the entire movie, although Adrian, her ex-husband, is an ever-present threat. Cecilia is prominently featured in the movie

---

[42] And some actually do. Lorena Bobbitt was made famous (or infamous, depending on your opinion) for cutting off her abusive husband's penis and tossing it out of a moving vehicle into a field in 1993.

posters, with minimal encroachment from the invisible man beyond a handprint or a spooky shadow. This is different from other horror movies, particularly slashers, where the villain features just as prominently, if not more so, than the protagonists, especially the female ones. The *Halloween* movies have been continually guilty of this.

In the posters for *The Invisible* Man the male threat, kind of like the social issues plaguing women, is invisible, threatening, and always there behind the haggard and scared faces of women. The gaslighting of the female protagonist in this film is not necessarily a method to move the plot forward, as Adrian likely would have continued harassing Cecilia even if she had been believed, as most abusers do. Instead, this gaslighting better reflects the real-life experiences of other women who are not believed.

Hollow Man (2000)[43] is another version of the original *The Invisible Man* except that it follows the original more closely: the main character, Sebastian Caine, is the villain and he torments the other people around him. It is rife with boobs[44] and the main character assaulting women. The first thing he does as the invisible man is take off his co-worker Sarah's sweater and feel her up while she is sleeping. When she wakes up, he pretends like he was never near her. He makes Janice afraid to go to bathroom, for fear that he is watching her, and he moves Linda's Coke can around on her to

---

[43] This movie has some of the most impressive CGI transitions of visible to invisible and back that I've ever seen. The movie as a whole is good, but even if it wasn't, I would say to watch the first half an hour just for those awesome transitions.

[44] You do get to see Kevin Bacon's butt and his penis several times, so maybe it balances out?

confuse her. He then rapes a woman, simply because he can, and proceeds to make everyone uncomfortable and disturbed by the fact that he could be around when they can't prove it. He is the ultimate gas lighter, using his ability to confuse and upset the people around him for his own psychological and mental enjoyment.

One thing that is different in real life as opposed to horror movies is that when a woman finally sheds light on abusers in real life, she is often questioned as to why it took her so long to talk about it. This statement ignores centuries of women being disbelieved by men and shunned for their accusations. For example, historically, for a woman to prove she was raped, she needed witnesses to corroborate her accusation (Leotta, 2018). The statement also ignores the decades of research on trauma and how victims or survivors respond to it, especially in cases of domestic violence or sexual assault.

Horror movies skip over that part and go straight into figuring out how to get rid of the killer. Part of this might be because most of the other characters are already dead, so there isn't anyone to say "I told you so" to, but maybe there is a deeper meaning to it. While I'm not typically one to dive too deeply into what a story must mean because everyone's interpretations are liable to be different, I like this one and thinks it holds merit: perhaps the reason that these women, who have been gaslit and mocked for the majority of the movie by friends, family, and law enforcement alike, are believed is that the horror genre feels they deserve to be believed. It is turning the rampant disbelief that plagues women and victims of crimes on its head and saying, "No. When people say bad things are happening, you need to believe them." It's not perfect, because you still usually have the disbelief beforehand. But, unlike when women bring forward hard evidence in real life against their abusers, there is no one seeking to excuse, mitigate, or explain away the villain's behavior. She is typically believed, without question, by the second or third act of the movie and she gets help.

Maybe the real world could learn a thing or two from horror movies.

Is That a Knife or Are You Just Happy to See Me?

"Sometimes all you need is a positive attitude and a knife."

In an episode of the podcast The Faculty of Horror hosted by Alexandra West and Andrea Subisatti (Episode 54: Undead Walking: *Night of the Living Dead* (1968), *Dawn of the Dead* (1978) and *Day of the Dead* (1985)[45] to be exact), Alex and Andrea comment on many of the social and cultural issues present in these films, and while some of their commentary was pretty obvious I nevertheless enjoyed their appraisals of the films. However, I have a major issue with what seems to be a throwaway comment on their part. In the episode, Alex and Andrea comment on how George Romero denies that he was trying to make a statement when casting Duane Jones as Ben. What is most striking to me is that they said they didn't care what the director intended because it was so brilliant he should take credit for it.

---

[45] The lack of an Oxford comma between *Dawn of the Dead* (1978) and *Day of the Dead* (1985) kills me.

Excuse me for a moment while I add some information to round out why this opinion[46] might be disrespectful and inaccurate. The part in question had originally been written for a crude, white male trucker and when Jones was cast simply for being the best audition for the part, Romero rewrote the dialogue at the behest of Jones, who was well educated, so that the part aligned more with Jones rather than the original idea for the role. Jones did refuse Romero's idea to rewrite the script so that Jones, the only black character in the film (and also the most likely to die), would not be killed at end. Jones felt that his death would be symbolically important and much more meaningful of an ending than if he lived. And boy was he right. Ben's role has led to people championing this movie as a commentary on race relations and racism. Viewers and critics alike attach much significance to this film and its contents, spawning books, studies, and commentaries for decades.

But what about Romero's opinion? Does that count for nothing? According to some people, it doesn't, even though it should.

When I was in high school, we had required reading that we needed to analyze for state exams and then write essays of these analyses which were then graded. All of the books we were to analyze were written by dead people. Why, though? Well, there was a story that a student once wrote a paper for which they received a failing mark due to not interpreting the text "correctly". This student wrote to the still living author who agreed with the student and told the school they were wrong. This led all schools in that district to

---

[46] And yes, it *is* just an opinion, but this all plays into the larger point of this chapter. Give it some time.

remove texts written by still living authors so that this could not happen again, and the teachers would be the final arbiters about what were, and were not, acceptable interpretations of books.

My point is that it's great that people have their own interpretations, but completely disregarding someone's purpose behind their own work denies that person ownership of what they have created. It's like saying that someone being sad about something doesn't matter because other people feel differently about that same thing.

We all look at things through our own lenses of lived experiences and thus will usually see things slightly differently than others. The view of the person who created the work is just as important though when considering one's own opinions and feelings about something. The problem with prioritizing one's own opinions is that sometimes things get a little out of hand with interpretations that fail to acknowledge the original intention behind a work. Or interpretations that fail to recognize that sometimes a good cigar is just a good cigar.[47]

The title of this section hearkens specifically to this message. Do people find things in movies and films because they are expecting to find them or because they want to find them? I noticed in a lot of my readings of articles, books, reviews, and other such commentary that the knife is frequently referred to as a large phallic representation of men violently penetrating women. Critics and researchers have expounded on the idea that the knife is chosen by

---

[47] There is no concrete evidence that Sigmund Freud ever said this, and this statement goes directly against everything that he ever studied, wrote, or psychoanalyzed. Yet this quote persistently follows his legacy.

killers because of this phallic representation, that they are using their penis and thus their masculinity to torture and kill women. Then, as according to Leslie Vernon in *Behind the Mask: The Rise of Leslie Vernon* (2006), when women pick up the "big, long, hard weapon" it is "deeply symbolic" of women either becoming more masculine or turning men's power against them. And, of course, this wouldn't be done with a "dinky little gun".

The movie I'm quoting from is a mockumentary horror comedy slasher, so it's meant to be taken tongue in cheek. However, many people subscribe to the beliefs that the knife is somehow a symbol of men's rage against women and women turning that rage against them. They believe it must be, or they want it to be, and that's all that matters to them.

But, when you really think about it, as a regular person and not as someone trying to find meaning in a film (which may not even be there), a knife, or any long sharp object, is a pretty desirable object to murder people with in horror films. Knives are easy to find, easy to get, and don't raise much suspicion if you buy one or are carrying one. The presence of a knife can usually be easily explained away, especially in a horror movie. Knives can be easily concealed, they do not make noise when you kill someone, and you don't need much experience or knowledge to know how to plunge the knife into a person's soft parts to kill them. Knives can come in a variety of sizes to better help the killer achieve their goals. Want a longer reach? Try a machete. Want something that will slide out of your sleeve? Try a kitchen knife. If the knife really was a symbolic representation of a penis, it would seem to me that the size of the knife would matter a lot more than it does in these movies. And, as in the case of some killers, when they do not penetrate the body and instead cut the victim's throat open or chop them in half, is the sexuality of the knife lost when it is used in that fashion?

When little Michael Myers kills his sister by stabbing her repeatedly with a knife are we to believe that he is harboring sexual fantasies about penetrating his sister? This would require that we sexualize a six-year-old boy which is wildly inappropriate (though perhaps only as bad as Bob suggesting to Annie that they tear off Lindsey's clothing after tearing off each other's clothes). Is Michael, and every other male horror villain, attempting to confront some sort of homosexual urges or conflicts by penetrating other men with knives and long sharp objects? Was Mrs. Voorhees expressing repressed sexual urges when she penetrated Jack's throat with an arrow, stabbed Steve and Bill, and repeatedly attempted to stab Alice?

This all just seems like too much of a stretch when the easier explanation would be that knives are readily available and easy to kill with.

Guns would make things easier from the viewpoint that you are more likely going to be able to kill someone with a gun than a knife. But guns require some amount of training and understanding in order to load, aim, and fire one. Depending on the gun, it may require some amount of strength to wield. How many times have we laughed while watching real people and actors in movies get blown backwards, or lose control of the gun, while firing it? Guns are more expensive and harder to come by. Guns will also penetrate a person, much in the same way that a knife will.

It is odd that, if a murder weapon is meant to represent a phallus and masculinity, then why would the gun not be the weapon of choice? Many people will accuse men of compensating for a small penis because they have a gun, as though the gun is an extension of their masculinity and a replacement for their penis. As mentioned before, a firearm will penetrate a person via a bullet, and a firearm even mimics a penis more than a knife as the gun will "ejaculate" a bullet at the point of excitement (when the person actually pulls the trigger). Does all this sound silly to you? Well, it doesn't sound any more silly than suggesting that a knife would be a phallus replacement in horror movies.

I could go on listing weapons (fire, rope, garrotes) and ways to die (strangulation, drowning, immolation, bludgeoning) like I'm trying to set up the plot of the next *Saw* movie, but that would be gratuitous. What does this mean for women who are so often penetrated by knife wielding villains in horror movies? Does this mean that they are, or are not, being subjected to male hostility through a symbolic phallus?

Don't get mad when I say that it is actually up to interpretation. It's almost impossible for it not to be! Unlike tangible statistics that show that women are not more likely to die in horror movies, or that black characters *are* more likely to die in horror movies (even if they *aren't* more likely to die first) (Barone, 2013), there aren't any available statistics about directors saying, "Yes, of course I meant for that knife to be like a penis. It is a symbolic representation of how men take advantage of women." Which is why, most of the time, it's up to people to determine how they perceive what they are viewing. In a survey conducted on

116

Survey Planet, 77 participants answered the question "Do you notice social issues commentary embedded in horror movies?" 57.1% of participants (44) said "Yes" and 13% (10) said "No". The remaining 29.9% of participants indicated that they didn't notice and they didn't care (11.7% or 9 participants); that they did notice and they didn't care (11.7% or 9 participants); or that they only noticed when it was an issue they cared about (6.5% or 5 participants).

Film has been an important vehicle to convey opinions about things pretty much since its inception, and horror films are frequently used to subtly (or not so subtly) highlight current problems happening in the zeitgeist they exist in. Horror films often push boundaries, so it would make sense for directors and writers to include other commentary that may make people uncomfortable or make them think. Most of the participants who responded to the poll (75.5%) said they noticed social issues, so whatever messaging is included in the films is reaching the audience, whether they care about it or not.

As important as a person's perception of the artistic material is, though, it is also important for viewers to acknowledge and respect the opinions and intentions of a creator as well. As much as the film can and will mean something to viewers, and may take on a life of its own, it is also meaningful to its creator and the literal horde of people who took part in bringing the movie to life. By denying their viewpoints, you are essentially saying that their motivation for making the movie is meaningless.

Instead, take your opinions as grains of salt and use them to season your open conversations with others. You'll probably find a new way of looking at things that doesn't involve turning anything that even remotely looks long and hard into a penis.

The Tangents

Show Me Your O (Shit) Face

"Perhaps it was pleasure that she'd heard. An orgasmic whoop, instead of the terror she'd taken it for. It was an easy mistake to make."

"The Hellbound Heart", Clive Barker

Have you ever been watching a horror movie and everything is going fine until the woman starts to scream in horror and it suddenly sounds a lot like...something else? Like you can't tell if she's enjoying what's happening or not?

Laurie Strode could definitely relate, as she confused Linda's screams during her strangulation for Annie squealing during sex.

You and Laurie are not alone. Humans actually do have a difficult time differentiating between screams of terror and screams of joy. In a research study conducted at Emory University in Atlanta, GA, it was found that when people heard screams of excited happiness without any context, they tended to think that the emotion being expressed was fear. The researchers hypothesized that this might be due to an evolutionary need to err on the side of caution: if screams of fear and joy both use similar acoustic features, you're going to want to assume the worst to be prepared for an attack. It was also found that listeners could not discern between real screams or acted screams, and that it was much more difficult to find recordings of happy screams for men. The researchers hypothesized that this is possibly because men don't frequently have happy exultations the same way women do because of the cultural expectations of how men are supposed to act when elated or ecstatic (Avery, 2021).

Beyond being an interesting bit of trivia to share with your next date, how does this relate to horror movies? Well, if we're looking at this from the idea that horror movies are made with men in mind, for a male gaze and male viewers who want to revel in and identify with a woman's pain and suffering in horror, it seems like an important bit of information. If one can confuse screams of terror for screams of ecstasy, what is the meaning that is being set forth in these movies? If the point is male eroticization of female suffering, then this would make a lot of sense. However, a lot of this book has made the point that most horror movies and directors are not really setting out to glorify female suffering, even if that is often identified as the end result by critics and researchers. Is it really the fault of directors, writers, and producers if humans cannot discern the difference between screams and naturally conflate happiness with terror? That would be equivalent to yelling at your child for slamming doors in anger after they have seen you do the same thing. They are acting out their basic instincts to imitate their parents, the same way humans are following generations of learned experience that it is better to err on the side of caution when hearing screams.

I opened up this tangent with a quote from Clive Barker's "The Hellbound Heart", the book on which his cult classic film *Hellraiser* (1987) was based. In the book as in the movies, the Cenobites combine pain and pleasure: either they provide you with so much pleasure that it becomes painful, or they just inflict pain on you in the guise of providing you pleasure as is their belief. They hear their victims' screams of fear or pain and acknowledge that for what it is while also telling the victims they should be getting pleasure from it. Sort of like an incredibly extreme, non-consensual Bondage, Discipline, Sadism, and Masochism (BDSM) experience [48] that you never get to leave.

The Cenobites take the original idea of not being able to differentiate between screams to the extreme. So just make sure you stay away from any weird puzzle boxes, otherwise you may be forced to experience the thin line between pain and pleasure for all eternity.

---

[48] If you're going to participate in BDSM definitely make sure that it is in a safe environment, with someone you trust, and that it is consensual. Nothing is less sexy than not having a freely consenting and happy partner.

What About the Female Gaze?

"They made Ripley a woman, without making her this helpless creature...I think I was very lucky. These were men who were creating this woman character, but they liked and respected strong women."

Sigourney Weaver

If the male gaze is all about the objectification of women, then the female gaze must be all about the objectification of men, right? That's usually how opposites work. People would probably be quick to point out the *Magic Mike* series[49] was directed by a...man (Steven Soderbergh). Yes, all three *Magic Mike* films were directed by the same man. So, what does this mean for the gaze being showcased in the film? Is it a male gaze because the director is a man? Or is it a female gaze because he's showcasing what women want to see (or at least what he thinks they want to see)?

Before actually defining what the female gaze is, I'd like to point out that the female gaze is not only available to women creators[50]. Men can also create through the lens of the female gaze,

---

[49] Three *Magic Mike* films but still no sequel to *Push* (2009).

it just doesn't happen as frequently as when women create something.

The female gaze can be an active and objectifying gaze which can be centered on both men and women, whereas the male gaze focuses solely on women as the objects of the gaze. The female gaze is more typically identified as an emotional and intimate view of people, seeing them not as objects but as functional parts of the society within the film. The people in the films are complex and have backstories. They cannot be replaced with sexy lamps and have the story still make sense.

Oddly enough, as much as horror films are supposedly filmed through the male gaze with the torture and objectification of women for the supposed pleasure of men, very few of the women in horror films can be replaced with sexy lamps and still maintain the structure of the story. However, even though those women cannot be replaced by lamps, they are still frequently on display as objects and their backstories can be shallow to non-existent: they are ultimately cannon fodder for the killers. Under a female gaze, even in horror movies saturated in blood and violence, there is emphasis on the characters and their emotions.

In *Jennifer's Body* (2009), directed by Karyn Kusama and written by Diablo Cody, the focus is not actually on Jennifer's body as the title might suggest. Instead, the film focuses on the relationships between Jennifer and Needy. To a smaller extent, we see the relationship between Needy and her mother, as well as how Needy and Jennifer interact with the male student body around them,

---

[50] The male gaze is not solely for men either, it is just rarely used by women.

including Needy's boyfriend, Chip. While Jennifer tears her way through boys every time she needs a new fix, the audience is privy to the rise and fall of Needy and Jennifer's relationship which plays a prominent role throughout the film and affects the way Needy interacts with other people in her life. Although Needy has a love interest, that is not the point of the movie and it is not the driving force behind the movie.

Promising Young Woman (2020) directed by Emerald Fennell is not your typical rape-revenge film in that Cassie has more depth to her character and backstory than just some vague throwaway lines about why she is in the situation that would lead her to take revenge against the men around her. She has a family life that affects her behaviors, she has relationships with others, and she grows and regresses as the movie plays itself out. In rape-revenge films under the male gaze, women are one dimensional plot devices, going from victim to marauding killer in ninety minutes.

The female gaze in women directed films eschews the fragmentation of women's bodies. Even when they are murdered or raped our focus is less on the violence being imparted on their bodies and more on their facial expressions, conveying the woman's emotions about the situation. I'll talk about that more in the chapter "Finding Catharsis in Strange Places", but it's important to identify here that in scenes where women are being hurt, female directors tend to keep their camera focused on the woman's face, forcing the audience to identify with her emotions rather than what is being done to her[51]. The audience is meant to identify with her and become

---

[51] This concept was referenced earlier when talking about Eli Roth's *The Green Inferno* (2013) when Justine is about to be mutilated.

uncomfortable with what is happening because they are aware that they are engaging with a person's pain, rather than just observing what is happening on screen. When Cassie watches the rape of her friend on her cellphone, the audience never sees the actual rape. They only see the growing horror and pain on Cassie's face as she watches what happened to her friend. Had this been filmed by a man, the audience likely would have seen grainy cell phone video of a dark room of men raping a faceless body.

Part of the female gaze is also meant to humanize male characters and take them out of the typical male roles and body types that the male gaze puts them in. In *Jennifer's Body*, none of the male characters that Needy and Jennifer interact with are stoic, muscle bound alpha male types. All of them, including the jock that Jennifer kills, are sensitive and have feelings.

With the emphasis on characters' feelings and who they are as people, there is less of an emphasis on their bodies, and this includes men's bodies. Men might think women want muscle bound alpha types, but those are the types of men that tend to be portrayed in movies directed by men under the male gaze (like *Magic Mike* or superhero movies). In reality, women tend to want to be around men who are able to express their emotions and whom they can feel safe around. Men have become victims of their own gaze, having bought into the male constructed idea that the "perfect" male body must have a lot of muscles and he must always be the coolest and strongest person in the room. Men may lament the unrealistic body expectations of ripped abs and huge biceps, but they perpetuate that male gaze.

In Kimberly Pierce's 2013 remake of *Carrie*, Margaret White, Carrie's mother, gets a more sympathetic background to add depth to her character rather than just blanket monstrosity. Carrie in Pierce's remake has greater control over her powers and uses them to target only the people that have harmed her. Stephen King's original novel and the first movie adaptation were meant to symbolize the lack of control young women have over their bodies and the fear they experience with these changes. The female gaze of Pierce's remake instead sees a young woman take control over what is happening to her.

As much as I support films that employ the female gaze, I find that I often didn't enjoy those horror movies the first time I saw them. Usually, if I watch it again or think about the movie, I find that I like it. This happened with *American Psycho* (2000)[52], *Jennifer's Body,* and *Promising Young Woman*, just to name a few. Others, like *The Babadook* (2014), I couldn't stomach to watch again because I found the characters just made me angry.

Why didn't I initially, or at all, like films directed with a female gaze? Shouldn't they have appealed more to me?

Most of the movies I didn't even know were directed by women when I saw them, so the argument can't be made that I would have gone into the movie already feeling biased one way or another about it. Taking a good long look at the way I felt about them, I identified that it was the *realness* of the characters that I

---

[52] You're probably wondering how this is shot from the female gaze when Patrick Bateman abuses and murders women throughout it. But, the focus rarely shows the violence and is more on his vanity and the toxic masculinity and narcissism that he exhibits which leads to him committing murder. This could be a whole chapter in itself.

initially disliked. These movies weren't shallow slashers that left you satisfied but unquestioning at the end of them. These characters tended to have an unexpected depth to them that turned me off, but still somehow drew me back in. Even *The Babadook* comes to mind every so often when I think about that great ending. I might even have liked the movie if I didn't despise how much the child screams and generally behaves badly as he acts out his confusion and grief.

For some reason, these movies didn't do it for me until I really sat with them and figured out my feelings for them. Which is the point of the female gaze: you're not consuming media simply for the sake of consuming media. You're consuming the media to get something out of it.

Think about it. Who are Hallmark movies designed for? Women. Although the storylines tend to be predictably cheesy and love always wins in the end, they continue to be popular because behind all the cheese and colorfully cheery color palettes, there are emotions to be felt, thoughts to be had about how our regular lives may mirror the Hallmark lives of these happy people. Just like how Oxygen and Discovery ID have capitalized on women's desire for all things murder and monstrous, Hallmark feeds other desires for security and happiness.

One of the big differences between these women centered Hallmark movies and female gaze horror movies is that you expect to have to confront difficult or familiar storylines in Hallmark movies. You don't expect to think about your backstabbing high school friend or someone close to you who committed suicide while watching a horror movie. Horror isn't really expected to frighten you that way, but the female gaze is bringing that to life for viewers.

Female screen writers also present an interesting dilemma about the female gaze when their work is directed by male directors. Through whose gaze are we viewing the action and characters of the movie?

Two recent movies, 2021's *Malignant* and 2022's *M3gan* were both written by Akela Cooper and directed by James Wan and Gerard Johnston respectively. *Malignant* managed to tackle questions of mental health and abusive relationships, though this got lost in the sometimes overly campy humor and weirdly shoehorned love interest side plot. *M3gan* addressed grief, childhood attachment, and the ability of the protagonist, Gemma, to juggle managing a child and her high stakes career. Both movies fleshed these ideas out even while managing to be gory and frightening. *M3gan* in particular focused on issues that tend to affect mostly women, like parenting and choosing between a career and family. Was this because the screen writer was a woman? If a man had written these stories, where might the focus have gone instead?

It is hard to know exactly whose gaze we are looking through, and from whom that gaze is coming from since men and women can present with opposite gazes. For example, Leigh Whannel, who has worked with James Wan from when they first created *Saw* (2003), presents a female gaze in *The Invisible Man* (2020) remake. Elizabeth Moss' terror is made more real by the fact that women are frequently the targets of stalking behaviors and abusive partners. With the right technology, this could be a real-life scenario, unlike the supernaturally powered slaughters committed by Jason, Freddy, or Michael Myers.

It is even possible to have a woman naked and on display for the entirety of the film without objectifying her under a female gaze. In *The Autopsy of Jane Doe* (2016), the body of Jane Doe is naked and uncovered while Austin and Tommy perform an autopsy on her after she was found at the scene of a family massacre. Despite the fact that Jane Doe is completely vulnerable and displayed, there is never a sense that she is an object or a passive observer in her autopsy. She is instead a menacing force that clearly unnerves the two coroners who treat her respectfully even up to the last terrifying scenes. Written and directed entirely by men (Ian Goldberg, Richard Naing, and André Øvredal), the movie has effective backstories for the characters that evoke emotion. Tommy and Austin are not simply there to be slaughtered by a malevolent force, they are there for the viewer to identify with. These are all hallmarks of the female gaze.

If you were expecting this to be as much of an excoriation of the female gaze as the chapter on the male gaze, then this was surely a disappointment. It was not for lack of trying that I did not find much fault with the female gaze, as I (probably much like you) assumed that it would be the objectification of men, much like how the male gaze presents women. Instead, I found some hints as to why I may not be as fond of movies directed by women, and it's because they force me to confront difficult emotions and relationships whereas male gaze movies do not. It is also interesting to consider the male or female gaze is not limited to either men or women but can be used interchangeably. Some men work well with a female gaze, and some women like to employ a male gaze. The gazes may also work together or clash, depending on who is directing and who is writing the movie.

Horror movies with a female gaze will likely continue to face an uphill battle with the general horror movie audience. While critics tend to delight in movies that provide more depth of character than gore, the audience does not always agree, and I tend to be in that second club as well. But, as more writers and directors focus movies through a female gaze, we may find that there is no option but to confront those feelings we would rather keep buried six feet under.

The Tangents

Is it illegal to show a penis on

television?

"Ask me about my weiner!"
Sherman Schrader, *Accepted* (2006)

That seems like a really weird question, doesn't it? I mean, of course it's not illegal. Many movies and television series show male genitalia on screen. The only thing the MPAA frowns on is showing an erect penis, or one flopping up and down so it looks erect.[53] So why this question? It's kind of a funny story.

Whenever I watch a movie or television show, I am watching to see who is naked and who is not. How much of that person is naked? Is it full frontal or just a little bit showing? Some side boob, a little glimpse of the base of a penis? I've been doing

---

[53] Which is why it was okay for Jason Segal to flop his penis back and forth in *Forgetting Sarah Marshall* (2008).

this long before I started this project because of a question a boy asked in my ninth grade Biology class: Is it illegal to show a penis on television?

His question came up because we had just watched a very graphic video of a woman giving birth. Some men don't even want to watch their wives give birth (and are sometimes even discouraged from watching), so why were we watching this as ninth graders? I don't know, and at this point it's irrelevant for me. What's more important was the response of our teacher to the question about penises on TV: bafflement. He didn't know how to respond. He at first said something along the lines of "No, of course it's not illegal." So, the ninth-grade boy (he looked exactly like Carrot Top with glasses and was typically the class clown so this serious line of questioning had us all thrown) asked, "Then why do we only ever see women naked on TV? Why don't we see more penises on TV?" The teacher kind of just stared at us, mumbled some sort of non-response, and then moved on with the rest of the lesson.

I was left with the burning question though. Why don't we see more penises on visual media? I've seen more breasts and vulvas in my life through media than I ever thought I would have. While this can happen in all sorts of genres, it is particularly central to the horror genre. A friend of mine said, "It's not a horror movie if you don't see tits." And, yeah, that kind of holds up, at least for most of early slasher movies and it's a trend that continues in contemporary horror movies. I can only think of three horror movies [54]

---

[54] I'm sure there are more. I have obviously not seen every single horror movie, and I may have also forgotten movies where we got a glimpse of a penis.

where you see a penis: *The Green Inferno* (2013), *Hollow Man* (2000), and *28 Days Later* (2002). I considered adding Ti West's 2022 horror film *X,* but I feel like a silhouetted prosthetic penis doesn't really count.

 The Green Inferno shows a little bit of a character's penis (right before spider crawls onto it) and I was frankly surprised given Eli Roth's gratuitous showing of breasts in *Cabin Fever* (2002) but no penis[55]. *The Hollow Man* shows Kevin Bacon's penis a lot, especially for a 2000 film with a big star. Granted, breasts are shown as well, but I'd say we see Kevin Bacon naked (even though his penis is sometimes in a state of turning in/visible CGI-ness) probably about as much as, or more than, women are shown naked. The movie shocked me. The first time we're shown his penis I had to rewind the movie to make sure I had seen correctly, and then I checked the year of the film. Kevin Bacon has advocated to "free the bacon" since about 2015, saying that gratuitous female nudity was not fair to actresses, and also not fair to actors who want to be naked, too. While I agree that men should free the bacon, I'm not so sure that women are doing it because they want to, and instead feel like they have to in order to make it in the film industry.

 Danny Boyle's *28 Days Later* decided to take it further two years later with the movie's introduction of Jim played by Cillian Murphy. Jim is naked when we first see him, vulnerable and spread across a hospital bed in a way that always makes me think of Christ on the crucifix. The movie focuses on this image for a good while.

---

[55] I thought he was going to doubly surprise me by not showing breasts in an exploitative way in *The Green Inferno*, but alas, he disappointed me here.

There is no playing coy like, "Whoops, did you see it or didn't you? Was it really there?" It's an extremely interesting choice for the time where the only place you really saw a penis on visual media might have been on HBO[56], and certainly not in theatrically released movies. Boyle also doesn't have either of his main female characters expose themselves. The only time breasts are shown is when naked rage virus zombie women are running alongside other naked rage virus zombie men.

So, why don't we see more penises? While there has been an increase of penises in television and movies, women are still far more likely to be shown nude than men. So, is it because film and television are primarily male dominated and they don't want to see other naked men? Is it an issue for the male actors? Is there a belief that people won't want to go to see a movie that shows male nudity? Cameras are typically viewed as having a male gaze, since men are usually behind them, and men will get squeamish about seeing other penises. Maybe that's why the majority of penises seen on television are prosthetics. It's fake and the owner doesn't have to feel self-conscious about any shortcomings. Unlike women who have their bodies displayed to be ogled and ridiculed. Apparently, men still get special privileges even when they're supposed to be baring it all.

One of my survey questions asked, "Are women more likely than men to be shown naked in horror movies?" 97.3% said yes and 2.7% said no. In a follow-up question, "Is a woman being naked more likely to be sexual than a man being naked?", 75% said yes, 15.4% said no, and 10.5 percent opted for the "Other" option. Of the

---

[56] And apparently those weren't even real.

"other" responses, some respondents suggested that it really depended on the context of the movie or the scene. One person seemed to specifically reference Alfred Hitchcock's *Psycho*: "Really depends on context. A woman getting stabbed in the shower isn't, a woman that's going to do horizontal tango yes." Another respondent said: "I'd say in most cases it does. Very rarely do we see a man without pants on but it's very often in any movie we see women topless." The most interesting response was "No idea, I'm asexual so never notice those things, in fact the last movie I saw (Starship Troopers, 1997) apparently had a scandal for the nude coed shower scene, I only saw soldiers in it."

Definitive scientific data? Definitely not. But these responses came from a variety of age groups, countries, and experiences with horror films (as per survey responses), and sometimes the public's perception of something is more important than what the hard data says.

At any rate, I'll still be keeping a mental tally of how much bacon I get to see.

Finding Catharsis in Strange Places

"We make up horrors to help us cope with the real ones."

Stephen King

The first time I watched *The Last House on the Left* it was the 2009 remake. It was viewed with a group of my friends and was one of those films that was chosen after we walked to Blockbuster and spent a half hour deciding what to choose. We didn't choose based on the descriptions on the back of the boxes. The choice was made entirely on the front cover and the tagline "If bad people hurt someone you love how far would you go to hurt them back?" It sounded pretty rad. Who doesn't want to hurt someone after they've hurt your loved one? It's a fantasy people will usually entertain at some point or another.

What we sat through was nearly two hours of torture with an ending that was barely satisfying. Mari and Paige suffer torture, terror, and rape before Paige is killed and Mari is nearly killed, left to float down the river to her home where she left her parents. The remainder of the film sees Mari's parents exacting revenge against the four strangers while she lays catatonically on a couch. The best part of the film is when the father microwaves Krug's head because

139

it is inventive and because we've all tried putting weird things in the microwave before just to see what happens.

But that's it. Mari is not triumphant. She is brutalized for a good portion of the film and then retains no agency to act on her behalf and exact her own revenge. The film made us all uncomfortable and left us feeling dirty and disappointed for having watched it. I hoped never to see anything like this again, and also hoped that there weren't many other films like this.

Imagine my dismay, then, when I found out years later that the movie I had watched was a remake of the 1972 *The Last House on the Left* which manages, somehow, to be even worse. Not only do Mari's parents avenge her in this move too, Mari also *dies*. She doesn't even get to survive her ordeal. The torture of the two women goes beyond what happens in the remake, though the rape may be considered not as graphic, and there are several oddly placed scenes that come across as comedic. The fact that it would have to even be said that interjecting comedy during these scenes is inappropriate should be unnecessary, and yet here it is: a male director finding comedic value in the torture, rape, and murder of Mari and Paige.

Actually, given the whole #metoo movement and what women everywhere know and have experienced from some men, it probably isn't all that surprising.

As I have dug more into watching horror movies and exploring the available fare on the numerous streaming services available to me, I've found that these types of predatory and sadistic films are more prevalent than I would have thought. I have also learned that these types of films have a name: rape-revenge horror. They originally became popular in the 1970's as political commentary, expressing the rage and terror of the Vietnam War, the race riots, the Civil Rights movement, and the general societal upheaval in the United States. Directors chose to show all of this metaphorically through gritty, brutal displays of both physical and sexual violence. While these films are banned in several countries and do not always receive glowing reviews, some people, including feminists surprisingly, praise these films for not holding back in showing the horrific reality of sexual violence against women. The problem with this, though, is that these films never address the equally brutal aftereffects of sexual violence. Instead, they follow the formulaic story of torture and vengeance.

These films have a very predictable set-up of acts: first, the female character is brutally raped and left for dead; second, she rehabilitates herself and/or plans her revenge; the third, and supposedly satisfying act, is where she exacts her revenge on her assaulters for the last twenty minutes of the movie after we watched her suffer for an hour. But, I suppose that is a better outcome than what the two *The Last House on the Left* movies did.

The questions that burn in my mind every time one of these movies comes up are: Why are these movies made? What draws us to them? What drew *me* to them? I've watched my fair share of rape-revenge horror movies out of what usually feels like a sick sort of curiosity. I don't feel better about myself or about the woman who took revenge into her own hands. I feel disturbed that someone (usually a man) would make something like this and think it is okay because the woman gets her revenge in the end. No one would think to make a rape only film where the perpetrator gets away unscathed and the victim suffers the physical, psychological, and emotional

141

damage on her own. That would be way too much like reality for an audience to stomach.

When these girls and women exact their revenge, what do they become in the movie? In some ways, they become monstrous: some of the ways that they kill their tormentors are heinous, and yet we as the audience condone it because we know those are bad men. The woman has been transformed from the object to be used into a monster. Neither of those things is a person though. Neither of those things has a place in society. These women are not fully formed characters who suffer at the hands of men and then take back their lost personhood through revenge. They in fact seem to lose whatever personhood may have been left to them after their victimization. Jennifer, the victim-avenger of *I Spit on Your Grave: Déjà Vu* (2020) diminishes herself down to her sexual appeal to men in a radio interview in the movie. Her sex appeal is her only weapon, she says, and she uses it to take advantage of her attackers to kill them. Not only does this sound unlikely, but it also diminishes Jennifer's sense of accomplishment in taking revenge: she has no useful skills or abilities, so she will simply seduce her attackers because that's the only thing she knows how to do.

The movies in which the woman does not get to exact her revenge, but instead has a proxy-avenger, are even more frustrating. Not only is the woman victimized by men (every so often there is a woman, too), she must be rescued by men. Men get to play out a typical male fantasy of playing the violent hero, the one who defends a woman's honor. But not the honor of a woman he doesn't know; only fathers and boyfriends/husbands ever get to avenge the woman. In movies where this happens, there is no talk about what happened to the female victim, or how her life will be forever changed because of her assault. The male avengers focus on pain and torture, with very little conversation except to maybe elicit some information about where other rapists are. They seem, even, to forget that the whole reason they are seeking revenge is for the woman in their life who was brutalized as her name and character rarely appear again once the man goes on a rampage. In these cases, the woman's world is controlled at both ends by men in her life. The initial group treats her as a doll, something to be used, abused, and played with. The second group treats her as…a doll. Except they perceive their doll as something that has been damaged by others and they are mad about it. The woman, meanwhile, has no say in any of this. These stories are more about the men's journeys, rather than the woman's.

The true aftermath of rape is also never discussed or addressed in any of these movies. The characters may cry and exhibit some anxiety, but their overwhelming and overarching emotion and behavioral response is rage. They are often alone in their revenge, seeking out neither friends nor family to assist them, as if they don't have any of these relationships. The movie has reduced them to a one-dimensional caricature of a woman: she is first an object to be used, and second a wrathful monster meant to punish. If anything, the only thing the rape seems to do to these women is to make them more adept and resourceful, sometimes almost superhuman. She is suddenly an expert at setting traps, stalking people, and dispatching much larger men without so much as getting another scratch.

A recent movie that makes an attempt at subverting these typical stereotypes of the rape-revenge subgenre is 2020's *Promising*

Young Woman. First, it's directed by a woman, Emerald Fennell. While there have been other rape-revenge films directed by women, it is such a small minority that it is still worth mentioning. Second, the actual rape of Cassie's friend Nina is never shown. The word rape is never used throughout the film, nor are the words assault or sexual assault. Euphemisms are used to describe what happened to Nina, as well as the fact that she killed herself. Even the revealing video Cassie watches towards the end of the movie is only noise and voices, but never is Nina's voice heard on the video. There is nothing to sexualize or ogle; no one will be getting off on the rape of the victim. It was one of the things I really enjoyed about the movie. Third, the movie plays with the typical three-part structure of the rape-revenge film, instead building a five-part movie over nearly two hours. In these five parts Cassie drags herself through small moments of making rape-y men potentially rethink their future actions, struggle to interact with her parents, work at a coffee shop job, and find a romantic interest in an old school friend, Ryan. Part five is when the real revenge is finally enacted.

While the movie does well in bringing to light the issues women face navigating an unfriendly world after they have been raped and of the trauma that takes hold of the people around them, Nina and Cassie remain stunted to characters whose entire existence is based around Nina's rape. Cassie finds a brief moment of time where she was able to move on and have healthy relationships, but it doesn't last long and it fits awkwardly into the movie. Like Jennifer in *I Spit on Your Grave: Déjà Vu,* Cassie uses her body to lure men in to punish them, as if this is all she is capable of after once being smart enough to be in med school. Nina does not get to take revenge on anyone, and Cassie's revenge only manifests after Nina's rapist has killed Cassie (in an agonizingly long suffocation sequence as bad as some rape scenes). We get some satisfaction in knowing that these men's lives will be destroyed by what they have done, but it rings a little hollow. Cassie had been able to dupe hundreds of men without ever getting injured, a trope which is part of this subgenre, but then gets killed and finishes her revenge from beyond the grave with very real consequences that are not as fantastical as her prior ability to get away with threatening and frightening men. What should have been the best part of the movie (as it usually is in most of these movies), when the main rapist cries and begs and pleads for her not to hurt him, also feels out of place. We had just been through a period of intense reality with Cassie enjoying a good life and being a real person only to be thrown back into her having these sneaky, preternatural abilities.

Frustratingly, the movie does not straddle the line between real and unreal, but hops back and forth in its desire to defeat the tropes of the genre. It left me disappointed with the final product.

Rape-revenge films themselves always have a line to straddle. How much or how little of the rape should be shown? Enough to justify the woman's monstrous revenge, but not too much so as to seem exploitative, meaning are people able to get off to it like they would with pornography? Isn't using any amount of a person's suffering exploitative? Especially when the suffering serves no greater purpose, as could be argued about rape-revenge movies. Are extended fifteen, twenty, thirty-minute rape scenes necessary to get across the point that someone has been brutalized? In 2007's *The Brave One*, Erica Brain (played by Jodie Foster) and her fiancé David Kirmani are attacked and beaten (David is beaten to death) in Central Park. The entire sequence, from the time the assailants first appear to when Erica is knocked out, is only about two minutes. That doesn't detract from the violence Erica suffered (which notably did not include a rape), the feelings of terror, or the rage and anger she would experience as she proceeds to kill any man in New York City who would dare to threaten her again. *The Brave One* is not a rape-revenge horror film, but as a crime thriller it still carries some of those same elements without the need for gratuitous nudity and violence against a woman. The violence served its purpose, and it was over.

Should the perspective be shot from the rapist's point of view, or from the victim's point of view? Who is the director trying to align the audience with? The last one should seem obvious, as aligning oneself with the rapist should only bring disgust, and yet we rarely see the rape from the point of view of the woman. There are many third person shots roving over the woman's naked or half-naked body as she struggles and screams, and from the rapist's point of view as he pulls off her pants or looks down at her or thrusts into her. Who do these shots serve? Certainly not the woman whose body and suffering are being put on display for everyone watching and therefore vicariously participating. Watching directly from the woman's point of view where she is looking up or back at her rapist, or at the ground in front of her, are better able to put the audience in the woman's shoes and takes away the gross voyeuristic aspect of the rape.

So, what is it about this genre that makes men feel so entitled to film a uniquely female perspective on the subject?[57] There are not

146

many films (horror or otherwise) which deal with the sexual assault or rape of men and the stigma and shame that follows those experiences, which in some ways are worse than the stigma and shame women face. Is it perhaps men fear confronting their own weaknesses? Maybe it is easier and safer for men to project their fears onto women and put them through the torture and trauma.

Would these films be better served if they were being written and directed with women at the helm though? If men are trying to confront their own fears and weaknesses through these films, wouldn't a female gaze or lens which focuses on emotion be beneficial for them?

Within the past decade or two, more women have come forward to helm horror films within the rape-revenge subgenre with a decidedly different take. *Promising Young Woman*, as already discussed, does not revel in the rape of Nina at all, not ever even using the words "rape" or "sexual assault". In the 2017 French film *Revenge*, Coralie Fargeat also does not show the rape of Jen, which happens off screen while another man stares guiltily away. The violence against the men is excessive and unrealistic (in the style of Tarantino), which Fargeat says was her goal. She felt that showing such extremes of violence would be like a cathartic release.

Another recent woman helmed film is 2017's *MFA* directed by Natalia Leite. In this one, the rape is shown but the camera remains on Noelle's face the entire time, denying any sort of pornographic gratification or voyeurism for the viewer. Noelle's

---

[57] I've addressed in other chapters that men can also be victims of sexual assault and rape. However, the majority of rape-revenge films center around women and are written, directed, and produced by men.

suffering is what we are focusing on and what we should be
identifying with.

Films like *Promising Young Woman* and *MFA* focus on the
horror and pain that these women go through, as well as the fallout
of such an event. The rape isn't happening in a bubble where it is
only the survivor and her attackers whom she systematically
dispatches. These women deal with their family, friends, grief,
shame, and anger in a real-world landscape with consequences.
They are throwing the male gaze out the window and showing an
audience what these events look like from a woman's perspective.

As women take on a greater role in tackling the rape-revenge
subgenre, it shows perhaps that women are better able to address
trauma that affects them, while men choose instead to subvert their
fears and feelings, putting them onto women. This may be a
reflection of society as whole where men are not encouraged to share
their feelings as much as women and are often routinely told to "man
up". It is just unfortunate that some men choose to man up by
graphically portraying women in desperate and traumatizing
situations.

What is unexpected is that a greater number of women are
choosing to direct and be involved in these types of films and that
women often make up the majority of the viewership. They seek to
make the woman the priority of the film rather than a man. This
woman is a priority not because she is someone's daughter or wife or
girlfriend, a status which essentially rescinds her own personhood
and replaces it with an ownership of someone else. Instead, she is a
priority because she is a human being, and that in itself is worthy of
compassion and justice.

Regardless, there is no denying the popularity (if you can call it that) of the subgenre among both men and women. Men may indeed be watching for the heroic male fantasy or the glorification of violence or because the rape scenes are gratuitous with a better storyline than a porn movie. Women, however, might be watching for far different and more meaningful reasons. Perhaps they really are finding catharsis in not so strange places.

The Tangents

What's your favorite scary movie?

"Oh, you wanna play psycho killer? Can I be the helpless victim? Okay, let me see...No, please don't kill me, Mr. Ghost Face. I wanna be in the sequel!"

Tatum, *Scream* (1996)

Scream (1996) is an iconic movie. Love it or hate it, there is no denying that *Scream* had a serious impact on horror and popular culture as a whole. I mean, who kills off the biggest star of the film within the first ten minutes? I kept waiting for Drew Barrymore to come back to life even after she had been disemboweled. She could have been snatched up and plopped into any of the previous teen slasher horror movies as your sweet blonde Final Girl. Yet Wes Craven straight up slashed all of our expectations to ribbons.

Wes Craven gave viewers a film where horror existed as a genre, rather than the characters being oblivious to it, with this incredibly effective opening sequence. Not only are we reeling from the death of Casey who we thought we were meant to identify with, viewers are also being treated to a universe where these characters have no excuses for making typical dumb horror movie mistakes. Horror *exists* in this universe. Randy Meeks uses this to highlight the absurdity and predictability of horror movie plot lines. Except, Wes Craven doesn't follow all of those rules.

While I've argued that there have been other movies (*Black Christmas* (1974)) and other Final Girls (Ripley) who did not conform to these rules, *Scream* was the first movie to call this out and let viewers know that at least some people in Hollywood knew how formulaic and ridiculous horror could be. This would make sense in this instance as the previous years had seen a glut of sequels and bad horror that tried to capitalize on easy money-making formulas. This also meant that there was some realization that viewers did not want the typical Final Girl who adhered to Carol Clover's definition. Viewers wanted characters who were real, not virginal archetypes.

While Drew Barrymore's Casey probably fit the Final Girl mold, Neve Campbell's Sidney Prescott did not: she partied, she had sex, she had flaws. Yet she persevered and also broke the meta fourth wall, declaring "Not in my movie" before shooting the killer in the face. *Scream* is significant in the way that it openly subverted and mocked the Final Girl trope, showing that it was completely okay for women to not adhere to society's standards for who deserved to live or die.

When Reddit user Elmstreet1985 asked "Why do girls love the SCREAM movies so much?", users provided fitting answers:

BillyAndStuAreLovers · 2 yr. ago

Easy. It has two strong final girls that live throughout the whole series. Normally a Final Girl is killed off unceremoniously in the next movie. Sidney and Gale subvert that trope.

⬆ 12 ⬇   💬 Reply   ⬆ Share   ···

From user JNTA1234:

Not only did *Scream* pave the way for horror to revamp itself through female characters with more agency and control, it also allowed horror to change up its narrative game, inspiring prolific directors all the way up to Jordan Peele. While I'm not saying *Scream* necessarily deserves all the thanks, the next time the supposedly dead killer doesn't come back to life for one last predictable scare, you just might want to tip your hat to Wes Craven.

And, just for the record, my top six favorite scary movies are, as follows and in no particular order: *Alien* (1979), *Jaws* (1975), *Train to Busan* (2016), *Gremlins* (1984), *The Descent* (2007), and *28 Days Later* (2002).[58]

---

[58] It was harder to make this list than I thought it would be. I tried to only choose five, but I wasn't able to.

Why Do We Think Women Die More Than Men in Horror Movies?

"There's a monster outside my window. Can I have a

glass of water?"

Bo Hess, *Signs* (2002)

Critics of the horror genre frequently point to horror's abuse of women as a reason for why it is a bad genre. They claim that women die most frequently and in the goriest ways. There isn't a lot of empirical research on this topic, and not even much anecdotal evidence. My own anecdotal experience from the movies I've watched (which would clearly not be every single horror movie ever) would tell me that men are more likely to die, and in greater numbers, even if the women die more brutally and with more voyeurism than the men. But, why would critics and regular viewers alike think that more females than males die in horror movies?

One of the reasons could be that women are usually killed in a more spectacular, noticeable, or sexual way. Think of long chase scenes, screaming and flailing on the ground as blood spurts across the screen, and the terror of her final pleading moment. Men typically are not given such extended scenes of their death. They're also not usually killed in a state of undress or nudity. The fantastical and sexual nature of something will definitely help to make it stick more in viewers' minds, giving us an illusory correlation effect that makes viewers think things are correlated when they potentially have nothing to do with each other.

Women are the spectacle of the horror movie, even if they may not have the lasting cultural influence of the villains. I mean, nobody is wearing t-shirts with Laurie or Nancy on them, or dressing up as Sidney Prescott for Halloween. But people do put on the likeness of Michael, Freddy, Ghostface, and other notable horror villains. Even the Alien from Ridley Scott's same named movie is plastered on shirts and has been made into costumes. Though Ripley at least, one of the earlier standouts as a female character who bucked the trends in horror movies, is frequently on shirts or cosplayed at conventions.

Chivalry, and the idea of women being a protected class, could also play into this. Much like dogs or children being killed in movies, viewers may have a special sensitivity to seeing women tortured and murdered because they are seen as a group meant to be protected. Men are not afforded the same protections and are expected to die in the service of protecting others. This would make their deaths far less memorable or meaningful to the viewer. It is also directly opposite to the male-as-sadistic-viewer role that is often applied to male viewers. We are more aware of women in stressful situations (or movies) because we have been conditioned to be that way, not because we want to see them suffer.

Table I. Slasher Films Analyzed

Alice Sweet Alice	I Spit on Your Grave
American Gothic	Killer Workout
Anguish	Last House on the Left
April Fool's Day	New Year's Evil
Bloody Birthday	Nightmare on Elm Street I
Chainsaw Massacre II	Nightmare on Elm Street II
Cheerleader Camp	Nightmare on Elm Street III
Chopping Mall	Nightmare on Elm Street IV
Class Reunion Massacre	Prom Night
Color Me Blood Red	Psycho II
Death House	Psycho III
Dream No Evil	Rock and Roll Nightmare
Drive in Massacre	Silent Night, Deadly Night
Fatal Pulse	Silent Night, Deadly Night II
Friday the 13th I	Silent Scream
Friday 13th III	Slaughter High
Friday the 13th IV	Sleepaway Camp
Friday the 13th V	Sleepaway Camp II
Friday the 13th VI	Slumber Party Massacre
Frightmare	Slumber Party Massacre II
Girl's Nite	The Black Room
Halloween	The Burning
Halloween II	The Deadly Intruder
Happy Birthday to Me	To All a Good Night
Hell Night	Toolbox Murders
Hello Mary Lou	Unsane
I Dismember Mama	Friday 13th VII

The one study that I could find about gender and survival in slasher films was done in 1990 by Gloria Cowan and Margaret O'Brien. Five students (all of whom were female) watched fifty-four films with 474 victims. They chose traditional slasher films and did not include films with supernatural or nonhuman villains. There were many criteria that Cowan and O'Brien used to identify what determinants would signal if a character was likely to survive or not. They tracked things such as sexual behavior, revealing clothing, provocative behavior (which also included taunting the villain), whether sex occurred before or during an attack, and whether characters displayed masculine or feminine traits.

They found that while females were no more likely to victimized than men were (49% versus 51%), females were more likely to survive than men (19% versus 10%). They also discovered that females who were portrayed sexually were more likely to die, as were males who used sexual language or who displayed negative masculine traits such as being egotistical, dictatorial, or feminine.

While this is only one study, it does pull its data from the time period that many would consider the heyday of slasher films, when they would have most likely been killing all of the women out of misogynism. However, women and men are attacked equally, and more women survive. Although, as one would have expected, women who acted sexually or who were sexualized were more likely to die.

One part of the study that was unexpected was that men who used sexual language were more likely to die than their peers that did not use that language. Typically, men who speak sexually are viewed as more masculine, and based on how horror movies treat men, the more masculine men should have been the ones to survive. However, maybe the use of sexual language was the male version of revealing clothing or sexual activity.

A more recent, but less scientific gender distribution of deaths tally comes from the Dead Meat YouTube Channel hosted by James Janisse. He has several hundred videos on the channel showcasing horror movies and their kill counts. He breaks down how the characters die, as well as how many of the characters were men versus women (or other varied creatures, monsters, or indescribables). I watched 88 of the videos at random, picking ones where the kill counts did not include aliens, poop monsters, or things that weren't human (so, pretty much like what Cowan and O'Brien did). Of those 88 movies, only twelve killed more women than men, making that 13% of the movies, while 87% of the movies killed more men than women. This didn't surprise me, and apparently, it didn't surprise Janisse either as he made a comment during one of the YouTube videos suggesting that this is to be expected: "A gender imbalance we are very used to seeing by now" (in reference to more men than women dying).

There was one eye catching article written in HuffPost titled "White Women Are Victimized More Than Any Other Demographic In Horror Movies" (Duca, 2013). While it isn't scientific research, I thought there might be something useful to be learned. The actual content, though, was not what the title promised. Instead, the author writes about how white women are the first to die – by a whopping 52% compared to white men, black men, and black women – but there was nothing about whether they are more or less likely to die than other characters, or whether more of them are killed overall. Something that the article noted, and which is probably very evident to most viewers of horror movies, is that minority characters tend to not even be present in horror movies. The odds of a character dying first when they appear most frequently in a movie are obviously going to be higher than other characters that don't appear as frequently. What I thought would be a promising article turned out to just be clickbait.

It seems that viewer perception of who dies in horror movies and how frequently they die does not actually match up with the evidence that I could find. Yes, women (particularly white women) may be the first to victimized, but the genre is also saturated with them. At least from an anecdotal point of view, that makes sense. Janisse's Dead Meat videos showcase a more recent exploration of the question of who dies more frequently in horror movies, even if they don't go as in depth into the reasoning as Cowan and O'Brien did.

One of the reasons critics and viewers accuse horror movies of being misogynistic is because they believe women are victimized and killed far more frequently than men are. Although the evidence is limited, that point doesn't hold up. In Cowan and O'Brien's study, women were killed more if they were sexual or had sex, but were still killed less frequently than their male peers. It was surprising to learn that men who used sexual language were far more likely to die than their peers who did not use that kind of language: 33% of non-surviving males versus 4% of surviving males used sexual language. Although this is not as pervasive as the punishment of women for sexuality, it seems that men are also being punished in some way for being sexually perverse. Either way, it's still the men doing most of the dying in horror movies. They have a lot more to be afraid of when the villain comes out than the women do.

The Tangents

But What Does the Internet Think?

"The internet is where some people go to show their true intelligence; others, their hidden stupidity."

Criss Jami

While doing my research, it was inevitable that I would come across internet forums of various use and vileness. For example, this comment in response to the question "Why do more men get tortured and killed in movies than women? (Not being sexist, just read the text)":

 dannyfleming0604 · 7 yr. ago

Because of people who say 'wa wa wa, you're being sexist by torturing a woman all the time'. The world is too sensitive.

⊖ ⇧ 1 ⇩ ⮐ Reply ⬆ Share ···

The original commenter pointed out that in movies like *Jurassic Park* (1993)*, Deadpool* (2016), and the 2006 version of *The Texas Chainsaw Massacre* only men die or get tortured, and if a woman happened to die then the viewer only saw "the aftermath" and not the actual kill. I'm not sure if these are the only movies this Reddit user has ever seen, or if they are just not actually watching

161

movies, but a few of the other commentors saw the same problems
that you, the reader, likely see with this question:

mks2000 · 7 yr. ago

Because you're only listing modern films that are attempting to subvert the norm that is traditionally the opposite.
You're also cherry picking. There are four in the Jurassic Park franchise. Tell me what happens to Zorra.

⤴ 10 ⤵   💬 Reply   ⬆ Share   ⋯

It is perplexing that for so many internet viewers who
proclaim to be movie experts, they've missed so much of what other
people have had to say on the subject of who dies or suffers most in
movies. What isn't surprising is the anger they espouse at their
perceiving that they are being targeted by "sensitive" people.

On the other hand, another user asked why so many more
women survive horror movies than men. One user, large_marge_,
suggested that the reason women survive while men do not is
because having a man defeat the killer would not make viewers as
invested in the story as when a weaker and more vulnerable
character, i.e. a woman, faces down the monster. It is an interesting
concept and aligns with the ideas talked about in the chapter
"Gaslight the Women": there needs to be tension and a certain level
of high stakes to get an audience involved and invested in what is
happening to the character. If that means having a weaker character
persevere against insurmountable odds, then the job is apparently for
a woman. One hiccup in this commenter's theory would be their
postulation that horror movies are marketed towards the young male
audience, as that is who mainly watches horror, and that's why the
protagonist must be a weaker female. The stakes simply would not
be high enough for a character who's shoes they could wear. This

doesn't hold up to the research (see "Why Do Women Love to Scream?"), but it's still something to consider.

On another thread, one commenter made a point that is completely antithetical to everything the academic community believes about horror:

KicksButtson · 7 yr. ago

Not to mention that having a female protagonist in danger is more likely to cause the audience to feel empathy towards her plight and root for her to survive. Where as with a male protagonist they'd be expecting him to do something brave to fight back and be a hero, and that's not usually the type of movie you get with the horror genre.

⬆ 8 ⬇   💬 Reply   ⬆ Share   ⋯

While it is a sort of backhanded misogynistic comment, suggesting that women could not be perceived as brave and capable of fighting back, KicksButtson seems to be of the mind that the plight of women in horror is not because men want to see them suffer, but because being female will inspire empathy and a desire for her survival.

It is always thought provoking when the average person's ideas of what is happening in movies are completely different from that of academics. And given the anonymity of a Reddit message board, we can't really know the gender of these commenters (though some might be obvious given the context of their comment). Some commenters didn't seem to care much about what the gender of the character might be:

@JeremyScoggins12 9 months ago

If Jason was Jennifer we still would watch. If Freddy was Franny we still would watch.

👍 👎   Reply

In my survey of about 78 respondents, in response to the question, "If you are a male, do you feel like you are able to to identify with a female protagonist in a horror movie?", 21.3% of respondents chose "Yes" while 22.7% of respondents chose "I don't think about that sort of thing." 2.7% of respondents chose "No" and the other 53.3% of respondents were female. For these respondents, at least not consciously, gender does not appear to be something males are all that concerned about when it comes to movies.

And, simply because the internet is rife with funny and memorable comments, I present this one which was made in response to the question of why women survive in horror movies but men don't:

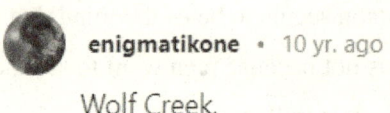

enigmatikone · 10 yr. ago

Wolf Creek.

⬆ 1 ⬇   💬 Reply   ⬆ Share   ⋯

Sad Girls Don't Die, They Become Villains

"No, I'm killing boys."

Jennifer's Body (2009)

Something that came up repeatedly in discussions about horror, trauma, and women taking back power that men try to take away from them is the concept and fruition of the woman villain. Not just old hags with an ax to grind or warty witches or abusive mothers (No wire hangers!) either. These girls and women may be conventionally hot and know how to use it or are super shy about it. They usually develop supernatural powers due to a trauma or an innate ability that manifests for any number of reasons. They are not the damsels in distress of the early monster horror movies or the slasher fodder destined for unique and overly long ways to die. They get to be the bad guy and they usually delight in it. For a time, at least.

There are three girls who popped up most frequently in this search: Jennifer in *Jennifer's Body* (2009), Carrie in *Carrie* (1976), and Ginger in *Ginger Snaps* (2000). As I write this out, I also notice

that all of these movies are named for their main character villain.
The title of *Jennifer's Body* would make one think the focus is on
Jennifer's body when it actually is not. Jennifer uses her body, sure,
but it's primarily to dupe men into getting things that she wants,
which later becomes the men's bodies as she needs to eat their flesh
to stay alive. The focus of the title and the movie though, is Jennifer
and not the male influences around her which she uses for her own
purposes.

The same thing applies for Carrie in the 1976 film[59]. The
movie is about her and how she inadvertently uses her powers to
exact revenge on her tormentors, teenage boys and girls alike. She
has very little control over what she does with her psychic abilities.
As for Ginger in *Ginger Snaps*, she has no control over her
transformation into a werewolf but uses her abilities to get back at
classmates she doesn't like. Carrie and Ginger are similar in that
their powers manifest around the time that they begin to menstruate,
which some people interpret as they become evil and scary because
they are becoming women who are threatening to men and therefore
must be destroyed.

Aside from the fact that Jennifer's transformation has
nothing to do with her period, I don't agree that Carrie and Ginger
becoming women has anything to do with their power or why they
are considered evil. Jennifer gains her powers after a failed satanic
ritual. The opening scenes of Carrie show her being bullied by other
kids because she has started her period and doesn't know what it is,
but I would argue that the real catalyst of why she develops her

---

[59] Kimberly Paige's 2013 remake gives Carrie significantly more control
over her abilities.

powers is because she has been bullied and tortured by the other kids in her school and has an abusive mother. She's not becoming a woman because she got her period for the first time; she's still a scared little girl who can't grow up because no one will give her the space to do it.

In *Ginger Snaps*, Ginger is probably the most responsible for her transformation to being a villain, although she was already pretty bad to start with anyway. She does, after all, get bitten on her way to steal another student's dog. The fact that she's on her period when this happens would play more into the myth that bears and other predators are more likely to attack women when they're on their period and that's why they should stay out of the woods. Her period does not grant her the powers which turn her into a werewolf, and it certainly did not make her evil. These girls do not become villains by becoming women. Had Carrie not been abused by her classmates and mother, her period would have come and gone without incident. Had Ginger not been prone to delinquent tendencies, she would not have been out where a werewolf would have attacked her for smelling her period blood, and her period also would have come and gone without incident.

All of these women also have their powers turned against them, which is not typical for male villains. Jennifer must consume the flesh of others in order to live. Carrie has no control over her powers and lashes out, ultimately becoming unstable and killing herself. Ginger also becomes the monster that her body forces her to be. All three of these women die, as villains are meant to, but it is at their own hands or by someone they love(d). Male horror movie villains are usually targeting strangers and are killed by strangers.

Male villains are (almost) always unambiguously bad. Not so with
these female villains.

Jennifer becomes a demon after a group of men take
advantage of her and try to sacrifice her to further their own goals.
Prior to going after a boy that Needy was interested in, Jennifer ate
boys who were interested in her only for her body and saw her as
nothing but a receptacle for their male teenage desires. There are
probably many girls and women who watched that part of the movie
with delight, feeling some slight vindication that the skeezy boys
who treat women as objects for their own pleasure were getting their
just desserts. Carrie's powers are also directed at the students who
torment her throughout the movie. Anyone who has been bullied
could have felt some satisfaction that those kids "got theirs" in the
end. Are Jennifer and Carrie really villains? Are they victims of
circumstance who were forced into their behaviors by things beyond
their control? Or are they just living out a taboo dream held by the
viewers?

Though I would not characterize Ginger as a victim of her
circumstances since she seems to greatly enjoy it, her werewolf
powers touch on another aspect that is usually used against women:
mental health. Ginger's behavior could be compared to someone in
a manic stage of bipolar. She is hypersexual and has unprotected
sex, she acts aggressively, she has little regard for those around her,
and she lacks control over her behaviors. Historically, women have
been called histrionic whenever they made valid complaints about
their lives. They were not believed about their concerns and treated
as children. Now, women's mental health in movies (and sometimes
in real life) is often represented as an oscillation between depression,

hysteria, and PMSing. She has no valid complaints, and it is all in her head. People sometimes perceive mental illness as the body turning against the person, which would also track with Jennifer, Ginger, and Carrie's bodies turning against themselves with their newfound powers.

These women are as much victims of their powers as they are villains with powers.

On the other hand, the movie *Teeth* (2007) presents another unwilling victim of her body who ends up using her powers for what some may consider good. In *Teeth*, Dawn doesn't quite understand the vagina dentata inside of her vagina. They have existed since she was very small (before her period, notably) and activate when men or boys penetrate her without her consent or when she is angry with them. Becoming a woman has nothing to do with her "ability" as it was always there, perhaps suggesting that women can protect themselves from birth, they just might not know how to do it.

Dawn initially fears them because she does not understand them, but once she realizes the power she wields, Dawn makes full use of it. She is not the villain in this movie, though she is the one killing and maiming others. The victimizing males are the clear villains here. While Dawn is viewed favorably, Jennifer and Carrie are treated as villains for making victims out of people who victimized them.

Critics and reviewers will sometimes write that female villains aren't scary because of their powers but instead are scary simply because they are women. Why would people want to deny women of their powers that make them stronger? Carrie, Jennifer, Ginger, and Dawn are not just women who strike fear into the hearts of men. They are women with powers beyond feminine wiles or muscles or sharp intellect. Whether it be a demonic force, psychic ability, or physical transformation, there is something that sets them way apart from other women. Many of the interpretations that these female villains are "just women" come from female writers which makes it seem as though they are trying to deny that the only way women can frighten men is through having powers. They want women to be scary to men simply for existing, which in many cases, is what men to do women. The number of men and boys I've seen on the internet talking about how women overreact in their fear of getting raped or assaulted, or who say they purposefully make women feel uncomfortable is appalling. Women are forced, in media and real life, to make space for men who denigrate them.

Yet women want to be feared, to have a measure of autonomy that allows them to walk the streets at night, to go home from a bar without having to deal with accusations of "She was asking for it" from men (and sometimes even other women). It is more desirable for a woman to feel like she can be scary and strong without powers because in reality she never will have powers like Carrier or Jennifer.

If it is not yet obvious, female villains are very different from male villains who do not have powers that turn against them or which they do not enjoy. The female monster is underestimated by the other characters, dismissed until her monstrosity can no longer be denied. She is more likely to look normal, only becoming monstrous in the act of her murders. Nobody dismisses Leatherface.

There is a category of women who have historically been feared by men and persecuted for those fears. Women who were not ordinary, or who did not fit the expected and demanded distinction of what it meant to be a woman. These women were burned at the stake, stoned, tortured, and drowned, whether they admitted to their sins or not. Today, they still drive concerns and fears amongst Christian and Catholic faiths, though the religious fervor does not quite reach the pitch it did centuries ago.

If you haven't figured it out yet, I'm talking about witches, a group of women who have never been underestimated, and in fact were irrationally feared. The horror genre didn't really start to deal with witches until the 1980's[60] and then there was a boom in the 1990's and 2000's of witch movies. There are still witch movies that come up now and then, and these more recent movies tend to portray the witches as victims, whether they have powers or not. 2015's *The Witch* sees Thomasin embrace the fear everyone feels for her and, like a self-fulfilling prophecy, becomes the witch everyone so feared she was. Thomasin wasn't a witch at the beginning of the movie. It was her treatment by society, and especially the male authority figures in her life, that drove her to evil.

However, the witch movie I want to focus on for this section is *The Autopsy of Jane Doe* (2016). Like Thomasin, Jane Doe in this movie is originally believed to be an innocent (by the characters in the movie at least – we all know something spooky is afoot). Initially during her autopsy, the coroners suggest that she may be a victim of human trafficking, an ultimate vulnerability for young women. Jane Doe is naked throughout the entire film. She is

---

[60] This isn't to say that there were not witchy films before this point. Just that they became more prominent at this point in horror history. *Rosemary's Baby* (1968) deals with witchcraft and devil worship, as does 1960's *Black Sunday*.

examined, poked, probed, opened up, and taken apart. As weird and progressively terrifying things happen in the funeral home, we, along with the coroners, discover that Jane Doe was (or is) a witch. Prior to her death she was brutally tortured and then burned at the stake, and now seems to be taking revenge on anyone who finds her. Even when the father, Tommy Tilden, tries to make amends and sacrifices himself to her, she remains rageful and violent.

Whether or not Jane Doe was a bad person prior to her mangling during the Salem Witch Trials is unknown, and I would argue that it is not relevant. The point isn't that Jane Doe is now a marauding psychopath supernaturally murdering anyone who finds her. Jane Doe was victimized at the hands of men, like many other women during the Salem Witch Trials who were all innocent (as we all now know from copious amounts of historical information). These women were targeted because they didn't fit the conventional norms of the time, were seen as a threat by prominent men in the community, or partook in spiritual practices that were not approved by Protestantism, like wicca or herbalism. Jane Doe feels so slighted by her treatment – and she rightfully should – that she does not care to forgive anyone or be choosy about who she kills. She is the absolute antithesis of what women are supposed to be or do when they are hurt by others. Women are meant to forgive people who hurt them or to accept their perpetrator's apologies, even if those apologies are meaningless. Jane Doe will do none of that, and because of that she is villainized.

By the end of the movie, viewers don't feel pity for her, as they may have originally when all of the torture she was put through is unearthed by Tommy and Austin. She is instead villainized because she refused to accept an apology and stop being angry. She is a rather unconventional female horror villain in that she is able to continue her maniacal massacre. No one can outwardly suspect her as she's dead, although the Sheriff seems to have a pretty good idea that she is bad news ("I want her out of my county."). We don't get a full, pitiable back story for her, and there are no real redeeming qualities to her. She's a female villain with the characteristics typically ascribed to male villains.

In Jordan Peele's movie *Get Out* (2017), he casts a young, beautiful white woman as a villain. She is not the "main" villain, if there really can be one in the movie, since the whole family is in on the body snatching scheme, but Rose Armitage is the most insidious and lasting villain. She lures Chris to her family's house and sets him up to have his body stolen. She acts like she is on his side until it is too late for him to "get out" (so punny). Rose even tries to hunt him down after he escapes, shooting at him with a rifle, reminiscent of slave catchers chasing after escaped slaves. While the film sets up larger, overarching theme villains (racism), Rose is the tangible villain, the one we can point at and say "That's her! She's the bad guy!" Peele doesn't just do amazing things with the disturbing and fearful nature of his film, he makes a believable and relevant female villain out of someone who is typically expected to be the Final Girl (white, beautiful, masculine in the way she confronts others for Chris). And, she's evil not because of some sad, traumatic backstory, either. She's just plain bad.

One of the earliest, and arguably most forgotten, female villains is Pamela Voorhees from the original *Friday the 13th*. Pamela, or Mrs. Voorhees, sets out on a murderous rampage spanning decades where she dupes people into believing she is a kindly woman who lost her son to drowning, when in fact she is the one murdering camp counselors, cops, and the camp owner. The entire time the audience is watching the film, Mrs. Voorhees is not even an afterthought. Everyone is thinking that it is, of course, Jason who is stalking and murdering everyone because that is what the formulaic slasher genre had conditioned them to believe. A deranged man is always the one to do the hacking and slashing.

Yet Pamela paints a far different and more interesting picture as a mother driven so mad by the death of her child that she sets out on a perpetual quest for vengeance against all camp counselors. Her rage doesn't necessarily imbue her with supernatural powers, as she struggles to fight against Alice, and she is summarily dispatched at the end of the film. Instead, she is sneaky about how and when she kills, and in some cases she turns people's underestimation of her against them.

Ursula, in *Thirteen Women* (1932) is an even earlier female slasher villain who is rarely, if ever, talked about or mentioned. It is possible that this is due to her using supernatural powers rather than physical powers, but many of the iconic male slashers have elements of the supernatural to them. Is there not enough blood? For the time period, there wasn't much blood to show: violence during the 1930's was largely implied and it certainly was not gory in the way viewers experience in the present day. Maybe the simplest answer – that it was a primarily female led movie with a minority[61] woman villain – is the reason why it's not heard of.

---

[61] To be clear, the actress who played Ursula , Myrna Loy, was not a minority, but for some reason was typically cast in "exotic" roles.

Kimberly Pinzon

Something to consider when looking at the relative dearth of female villains in horror films (or film in general) is the readily accepted notion that women just do not commit as many crimes, or as violent of crimes, as men do. Historically, women would more likely have been arrested for "minor" crimes like stealing and prostitution. The idea that a woman would or could commit a violent crime like murder was essentially unthought of because women just didn't kill others, unless they killed their children due to the "baby blues" or some sort of psychotic break. Lizzie Borden, who took an axe and gave her mother and father forty and forty-one whacks respectively, was acquitted of the murders. Some believe that was due to the very feminine way she presented herself in court, as opposed to the way she normally presented herself to others. The male dominated criminal justice system simply did not think women capable of the same type of violence as men.

Annie Wilkes, the number one fan of Paul Sheldon in *Misery* (1990)[62], is shocking in her role as kidnapper and torturer because she is a woman. Would the movie have been half as shocking and disturbing if Annie had been Arnold? Probably not. A man who kidnaps, drugs, and tortures someone into doing something that man wants could very easily read as tomorrow's newspaper headline. The same could not be said of a woman. Annie's motivations could be similar to a man's though: she idolizes Paul Sheldon and is angry when he kills off her favorite character in his popular book series. Becoming angry and violent because she didn't get her way is very male coded behavior, yet here it takes the form of unsuspecting

---

[62] The screen adaptation of Stephen King's book "Misery".

Annie Wilkes. Who, by the way, also killed multiple people when she worked as a nurse in a hospital which is something that women have done in real life as well.[63] Had Paul Sheldon not had such a considerable will to live, Annie would have gotten away with killing him as well.

The opinion that women are too fragile or gentle to commit crime has been proven wrong over the past few decades as what some people call "pink collar crime" has been on the rise. The gender gap in crime, where men commit significantly more crime in general, has been closing. Not only are more women getting arrested and incarcerated for crimes, but the inverse is happening for men, where fewer men are being arrested and incarcerated. While men still hold the highest numbers on violent crime and crime in general, researchers and other people interested in crime trends are trying to figure out why this inverse trend is happening (Chun, 2020).

Criminologist Freda Adler posited in 1975 that women's liberation would lead to an increase in crimes committed by women because more women would be put in positions of power and have more opportunities to commit crimes. Other crime analysts believe it is because men are starting to lose the "chivalrous" idea that women could not possibly commit crimes. Kelly Paxton, known as the Pink Collar Crime Lady, explains that because women are usually loved and trusted, people assume that they would not commit crimes. Essentially, being a woman is the perfect cover for being a criminal because no one would ever suspect you.

---

[63] Genene Jones, a real-life serial killer who, as a nurse, killed dozens of infants and children in the hospital during the 1970's and 1980's, was the inspiration for Annie Wilkes. Women who engage in this type of murder typically earn the moniker "Angel of Death".

Kimberly Pinzon

This all ties into why it is so incredibly shocking when a woman is the villain in a horror movie: we simply expect that villain to be a man. It is always a "twist" ending when it isn't the male character you expected murdering everyone, but the female character you wrote off as "most likely to die" who is doing the murdering.

Will there ever come a time when directors won't be able to rely on this relatively easy way to add a twist into their movies? Potentially, but I wouldn't hold my breath. Male villains still dominate the horror scene, for better or worse. As much as horror sets up women to be good Final Girls, triumphant survivors, or even traumatized victims, it has some work do to in making female villains as monstrous as male villains.

The Tangents

The Descent (2005) is a feminist horror masterpiece

"I'm an English teacher. Not fucking Tomb Raider."

Beth, *The Descent* (2005)

One of my favorite movies was originally one that I laughed through when I saw it in theaters. I saw it with my friend and my sister (there was only one other person in the theater) and I think we laughed our way through it as a defense mechanism to not be scared because, frankly, the movie is terrifying. Creepy sub-human goblins triggering the uncanny valley reflex hunting down a group of women one by one? Holy hell that is nightmare fuel. On top of that, you have the relationships between the six women that adds to the tension. One woman is wildly reckless, another is too inexperienced, and another has some dark secrets which drive her to do terrible things in the caves.

Man versus nature versus man plays out in ways that are unexpected as the women are picked off, not just by the monsters, but by each other. Each woman is distinct from the other personality wise, and the director even had each of them use a different accent to further distinguish them (I personally can't really tell the differences in the accents). In the beginning, any of them could be a potential Final Girl, though we all know it is likely to be Sarah or her best friend Juno, who seems like the more capable of the pair. But the fact that the possibility is there for several Final Girls instead of the obvious Final Girl pushed in most other horror movies is delightful. Neil Marshall doesn't waste much time introducing the first tragedy in the form of Sarah's husband and daughter dying, and we quickly move into the caves after a brief getting to know you montage for the women.

The cave is where everything starts to fall apart when Juno purposefully takes the group to a different cave than the one they were supposed to go to. Injuries happen, panic sets in, the women start to bicker the way groups do when put in high stress situations, and then the monsters come.

I don't consider this to be a feminist horror masterpiece because it's six women who bonded and then are put through the wringer. I consider it to be this because you could switch out every character with a man, and it would still make sense. There is nothing in the movie that screams "Look! These are women! Look how strong they are because they are WOMEN! Hear them roar!" Movies never have to show how strong men are, so the choices Neil Marshall uses to show these women's strengths and weaknesses without saturating everything in pink, frills, and bubbly mimosas is refreshing.

Marshall's movie differs from many other horror movies in which women feature prominently in that these women are not perfect. Even Final Girl Sarah Carter, who has the added sympathy and deference of having a dead husband and daughter, refuses to take the metaphorical high road and instead chooses vengeance. One of the pitfalls of the traditional Final Girl mold is that they are meant to do everything perfect so that when they survive, the viewer should feel as though they "deserve" to survive. While, in general, characters in horror movies survive based on their trope-centered personality characteristics, Marshall eschews all of that and writes real characters regardless of what is expected through horror tropes.

The flaws that each woman brings into the cave are also real and human. Mostly, they are scared, with the real flaws being Holly's impulsive rashness, Sarah's withdrawn and fragile personality, and Juno's continuous string of bad decisions. The women are interesting in their own rights through how they tackle the struggles and terrors of the cave. Marshall successfully made strong, savage women without the need for all of them to have a traumatic back story. And even though Sarah has some trauma in her backstory, it isn't what propels her savagery in the movie[64]. She is simply out to survive any way possible because she wants to live.

---

[64] Unless you believe the theory that there are no monsters in the cave and Sarah is the one killing her friends.

Why do Women Love to Scream?

"Girls don't want rom coms. We want murder."

My sister

The fact that women love horror continues to surprise directors, producers, reviewers, and male horror enthusiasts year after year. Somehow, even though women fill the movie theaters, rent the movies, and talk about what their favorite horror movies are, their love of horror still confounds. To a certain degree I get it. After all, even though women fill the ranks of the main protagonists in horror movies, they often go through a lot of suffering and the other non-main character women are often sexualized and killed off.

Yet women continually make up the majority of fans who attend the shows: 60% of people in the theater for *The Ring* (2002) were women; viewership for *The Grudge* (2004) was 65% female; and moviegoers for *The Exorcism of Emily Rose* (2005) were 51% female (Spines, 2009).

When FEARnet decided to launch a video-on-demand channel for horror movies focused on torture porn (the exploitative subgenre focused on extreme amounts of gore, pain, and torture, usually against women) their focus groups found that it was women who wanted the channel more than the men did (Spines, 2009).

The television channel Oxygen, which attempted to cater to the female demographic through smushy romance shows, tapped into this secret well of women who like all things disturbing and depraved. They started small with a dedicated true crime block in 2015 and saw a 42% increase in viewership. When Oxygen changed their format to all true crime, all the time, their viewership rose even more (Klein, 2019). Investigation Discovery (or Discovery ID), a channel devoted entirely to murder, has consistently ranked as the number one channel watched by women (Battaglio, 2016).

It seems pretty clear what women are interested in watching in their free time.

The lack of understanding appears to come from research that tries to conclude that women are not mentally or emotionally suited for horror films. After reviewing research on the psychology of horror writing in 2019, a professor named G. Neil Martin suggested that men and boys prefer horror movies more than women because men and boys have the low empathy and fearlessness needed to enjoy horror. Another doctor in 2020, Allison Forti, claimed that people who enjoy horror have lower levels of baseline anxiety. This would, according to Forti, exclude women from enjoying horror as women are nearly twice as likely as men to develop an anxiety disorder.

Despite what these doctors have concluded, other research shows that 43-60% of women enjoy horror (the discrepancy seems to come based on the year the data is from, and the specificity of the questions asked) (Rubin, 2018 and Civic Science, 2017). A CBS news poll from 2021 found that 48% of people surveyed enjoy horror movies, and if 50.5% of the United States population is female according to the 2020 census, it seems like a good portion of that 48% is likely to be women. I wish I could give you harder statistics, but when combined with the viewership statistics for *The Ring, The Grudge, The Exorcism of Emily Rose*, the FEARnet focus group information, Investigation Discovery, and Oxygen, I can confidently say that women just love horror.[65]

A non-scientific way to look at this is through the number of memes regarding how obsessed women are with horror and true crime. There are a multitude of women helmed podcasts regarding true crime and horror as well. If women were too anxiety ridden, too fearful, or too empathetic for horror, it would stand to reason that this would not be as big of a cultural phenomenon as it is.

Vera Farmiga (a horror fan herself who starred in *The Conjuring* series) suggested that because women are more emotional, they crave the visceral experiences that horror movies can provide (Spines, 2009). Horror pushes the boundaries of comfort and what the human mind can endure, often providing garish and startling visuals of violence, or stimulating us in more insidious psychological and emotional ways. Horror is one of the genres where it is difficult to watch it without feeling something, even if that feeling is disgust. Women are typically considered to be far more in touch with their feelings than men, so horror may provide a safe outlet for

---

[65] I include true crime channels here because true crime is horror made real. Even horror films that are "based on true events" are not as real as watching actual victims, detectives, and family members appear on screen and talk about the horrors they have experienced and witnessed. Even without jump scares, intense musical stingers, or mood lighting, true crime is horror in the flesh and, in my opinion, another gateway into understanding why women love to see other people (mostly women) suffer and die.

experiencing scary, painful, or disturbing feelings. This operates much in the same way that riding a roller coaster provides a safe way to experience an adrenaline rush or terror: you know that you're eventually going to get off the ride (unless you're in a *Final Destination* movie). These manufactured horrors have a specific end time, allowing women the safety to know that they will be scared but the scares will end after a pre-determined amount of time.

A reason that may resonate with women is this idea that, deep down, we enjoy watching a monster unleash murder and mayhem because women are so controlled and judged in everyday life. Women are subjected to numerous indignities, small and large, regarding their behaviors, looks, personalities, and anything else which may be commented on. They must act and be a certain way so as not to be subject to ridicule or judgement. What must it be like to act with such willful abandon as Michael Myers, Carrie, or a xenomorph? Something so freeing, even if devastating to others, is a fantasy women can safely live out vicariously through horror movies. Women, or viewers in general who enjoy horror movies, can put a safe psychological distance between themselves and the violence witnessed. They can identify with the feeling of freedom, but not embody the violence.

Horror may be more attractive to women because women are frequently the protagonists, leading characters, or majority of the cast in horror films. This has happened historically throughout most of the existence of the genre. Ratings on horror movies do not tend to fluctuate due to the nature of the female cast members either. Examples of this would be when people purposefully rated *Captain Marvel* (2019) poorly; the poor reception of the woman led

Ghostbusters (2016); or the poor ratings of *Ocean's 8* (2018). When a horror movie is bad, viewers do not turn on the women in the films, instead they say that the movie was too gory, too long, too boring, or too mindless. Somehow, the women in horror movies still maintain respect even if the film does not.

Women may be drawn to horror because it is a cathartic experience for them. Women can not only live vicariously through the villains, but they can also experience and re-experience traumatic events with a definite resolution. Violence and trauma in real life are messy and do not always have a clear cut or easy ending. Post-traumatic stress and other mental issues, along with financial, familial, and other life strains due to the violence or trauma can last for years. In a horror movie, these problems conclude at the end of the 90 to 120 minutes you spend with the character. The resolution is usually neat and final. This might be why some women choose to watch rape-revenge films: the trauma of sexual assault can get explored and have a triumphant ending all within a prescribed amount of time. Women can vicariously even the score with perpetrators from the safety of their own home.

Or perhaps it allows them to tap into the darker part of themselves that doesn't frequently get attention. Tying back into the earlier notion of women needing to adhere to stricter standards to be "acceptable" to society, women are expected to not be as violent or aggressive as men. When they are angry, women are expected to play the peacekeeper or acquiesce to other's demands in order to keep the peace. Whatever dark, wrathful, or aggressive thoughts a woman may hold, she is not allowed to express them. Watching horror may give women the permission that they need to tap into and feel those emotions in a socially acceptable way. I would almost guarantee that every woman who has watched horror has, at some point, thought that they could have done a better job at murdering and cleaning up all those dead teenager bodies.

Horror movies may prove to be appealing through their usage of tension, relevance, and their unrealism. Tension is an important part of any type of media that wants to keep you interested in it. Even Hallmark movies have a measure of tension: will they make it in time for Christmas and still love each other to show the true power of the holidays? You know the odds are exceptionally good that they will, but the possibility is there that it may not work out. There must be some kind of payoff through the tension to keep us invested. Horror movies do this well because the endings are not always what you expect. Sometimes characters you expected to live, bite the dust. Or the killer survives. As tropey as horror can be, it can still manage to pull some tricks out that give a viewer good reason to be tense.

Like with many other mediums, viewers want to find relevance in what they're watching. If it is not doing something for them psychologically, they are not going to keep going back to it. This is especially important in the case of horror movies where people can be turned off by violence, gore, or other disturbing imagery. If they can find a deeper meaning in it – whether personally, culturally, or some other significance – then they will be more invested. Or if the movie makes them feel strongly about it, viewers will remember that and be more likely to seek out other movies like it. Movies can also be considered relevant simply because of who they are directed by, or the content they contain.

Kimberly Pinzon

Movies by Jordan Peele, Ari Aster, or James Wan are immediately relevant to many horror consumers based on their previous experiences with those directors. Films that focus on fear of the unknown or giant monster spiders would be relevant for the population that has those fears.

Unrealism addresses the idea that we know that we will still be safe after the movie is over because we are watching it safely removed from the evil entities within. Despite special effects that can make the flaying alive of a person look incredibly real, we know that it is fake and that the actor is not actually being harmed. The same people who consume graphic and bloody horror movies are, according to research, still likely to feel disgust at actual representations of violence and are more likely to avoid watching that (Haidt, J., McCauley, C., & Rozin, P., 1994). Viewers want to watch pretend suffering and death, not actual suffering and death.

Regardless of the idea that the safety in unrealism plays a role in people's enjoyment of horror, I believe that this is probably the weakest link in this trio. Despite the remove from which viewers enjoy horror movies, the films still cause people to change their behaviors after watching them. People stop going into the water after watching *Jaws*; people refuse to play with Ouija boards (or sometimes go out of their way to play with them) after watching movies featuring the boards. Sleeping with the lights on, or keeping your feet under the covers, or being a little more cautious around cemeteries can be a result of watching horror. We know these movies are not real, but they still affect us.

Another problem with the unrealism aspect is that found footage films or films claiming to be based on real events are wildly

189

popular. *The Blair Witch Project* (1999) was the first found footage film to really make it big, and to use such a creative marketing scheme as to make people believe the whole thing was real. And people absolutely loved it. The movie was made on a budget of under $500,000 and grossed $248.6 million dollars in box office sales alone. That's 497 times what was spent on making the movie. It's also considered to be a cornerstone of the horror genre, being a springboard for the found footage subgenre to take off and reviving immersive marketing techniques. If we really want to be separate from scary things (like people who avoid the news do), then why would we seek this out?

What about movies "based on real events"? And I don't mean *The Texas Chainsaw Massacre* kind where it seems based on true events but is really just loosely based on an idea. Exorcism movies like *The Exorcism of Emily Rose* (2005), *The Conjuring* series, or the *Insidious* series tend to do very well, drawing in people year after year. But, for religious or spiritual people, the idea of exorcisms or possession by evil spirits is a very real and present fear. Seventy to eighty percent of Americans identify with one religion or another (Pew Research Center, 2022), and yet these movies do well. Home invasion movies like *The Strangers* (2008), *Hush* (2016), *When A Stranger Calls* (1979), *The Purge* (2013), or *Funny Games* (2007) seem to be popular (at least judging by the number of remakes and sequels) and yet these probably hit even closer to home for the average person than an exorcism movie might. Anywhere between 1.1 and 3.7 million burglaries happened each year between 2007 and 2020 according to FBI statistics (which are based on reports made to police and don't include times when people do not report) (Unified Crime Report, 2022). While burglaries and home invasions are different (no one is home during a burglary, whereas the occupants are present during a home invasion), the risk is still very real.

There are also instances where the unrealism factor fails entirely such as when people faint or vomit during movies. The tagline of the 1972 *The Last House on the Left* was "To avoid fainting keep repeating it's only a movie". Even when people know they are safe, or the movie is not based on true events, or has many elements of unrealism, they still have powerful, visceral reactions to what is being observed on the screen.

I think the biggest component here is that the movie makes people feel something. We don't typically engage in media because it makes us feel bad or because it makes us bored. Horror engages us in ways that create cult classics, loyal followings, and obsessions with characters who wouldn't or shouldn't be idolized in real life. It makes you feel scared in a "safe" way, even if you don't linger at the basement stairs or wander in the woods at night anymore.

We can screech in delight or disbelief "Don't open that door!" or "Why are you hiding there?!" Horror is an interactive experience that touches on our primal urges to congregate for safety and avoid things that look suspicious. Movies are usually better with friends, but none more so than horror movies. Horror is a community experience which would lend it towards being a woman's genre: women tend to be the ones who congregate and bring people together.

We can watch horror films and wonder how we would react to certain situations or decide which characters we do or do not want to die. Horror can help us look inside ourselves and find our own strengths or desires.

Okay, okay, I'm really waxing poetic on all the things I love about horror. But I can't help it. And I'm not sorry. I just love to scream.

Is Horror Really Misogynistic Trash?

"In all the horror films that I have done, all of those women were strong women. I don't feel I ever played the victim, although I was always in jeopardy."

Adrienne Barbeau

Although I am grateful for movements and organizations that bring the issues of women's rights, equal pay, sexual assault, and others to the forefront of cultural conversations, there is always a question of have we gone too far, or are we approaching that point?

One of the marketing advertisements for *X-Men Apocalypse* (2016) shows the villain Apocalypse choking Mystique, a shapeshifting female mutant. There was an immense outcry against this image, claiming that it promoted violence against women and was offensive. 20th Century Fox removed the ads and issued an apology for them.

People applauded this decision, but was it really the right one? Considering that one in four men have experienced some form of physical violence from an intimate partner, one in twenty-five men have admitted[66] to being injured by an intimate partner, and one

---

[66] This is an important word because violence against men in intimate partner relationships is often significantly underreported.

in eighteen men have been stalked by an intimate partner in their lifetime, it's clear that men can also be victims of violence. It can touch anyone regardless of gender. Yet I seriously doubt that anyone would have made an issue if Apocalypse had been strangling Wolverine or if Rogue had been strangling Cyclops.

In a superhero movie, villains do not discern between genders when inflicting violence and damage on the superheroes. Everyone is fair game. For people to claim that this advertisement somehow promotes violence against women would lead me to believe that they are also unaware that these women are also assaulted by the villains in the movie. They are enemies fighting against one another after all. Protesting violence against women in movies simply because the victim is a woman is dismissing who that woman is as a person and reducing her to only her socially designated gender label. Not very progressive or feminist of these people. And, though it is not as frequent, women do commit domestic violence offenses against men and commit general violence against men. Violence in movies should be equal opportunity.

The same could be said with horror movies which are constantly brought up in conversations and research about misogyny in films. Violence against all characters in horror movies is an essential part of the genre, otherwise it would be a completely different genre or type of movie. Sometimes this violence is purely circumstantial, sometimes it is necessary to move the plot forward, and sometimes, yes, the violence is performative and overly gory and explicit. The questions then become:

Is most of this extra violence directed towards women?

194

Is there more of a focus on women's pain and death than on men's pain and death?

If horror films are misogynistic, why are women almost always the ones to survive, and usually, defeat whatever evil is targeting her and her friends/family?

Tackling the last question first, it is undeniable that women are typically framed as survivors or "winners" in horror movies, especially slashers. This is the whole reason why the Final Girl trope exists! Regardless of some of the issues previously mentioned with this trope, the women survive to fight another day. "Fight" is definitely the correct word because there is a lot of trauma that is going to be following these women (and men in some cases) for the rest of their lives. But fight, survive, persevere, triumph, whatever word you would like to choose, these women embody it and set an example for other girls and women who are seeing and potentially looking up to them, and they're showing the boys and men who are watching that women can conquer, too. If you are going to try and make a case for horror movies being misogynistic, this is going to be a pretty big hurdle to overcome. Violence against women is typically perpetrated by men in real life and in horror movies. Perhaps the Final Girl character is way for women to vicariously fight back against the violence they experience.

Statistically, women and men are both targeted almost equally when it comes to violence in slasher movies based on a 1990 study done by Gloria Cowan and Margaret O'Brien. What was most surprising about this research was that the male characters who were most likely to survive demonstrated positive feminine qualities such as being empathic, expressing their feelings, and being less cynical.

Male characters who were viewed as more masculine, or who had negative masculine traits like cynicism or being dictatorial, were more likely to die. I find this particularly interesting given the idea that female protagonists in horror movies are frequently masculinized in order to supposedly make them more appealing to the male audience. Yet the male characters who survive are not the most masculine. It is as if there is some sort of male fantasy attempting to be lived out through these characters through the masculinization of the women and feminizing of the men. Maybe men are not trying to see masculine women. Maybe they are trying to find a middle ground against what society expects of them, and they get that through stronger women and more effeminate men in horror. This would also play into women identifying with the stronger female characters and the softer male characters: they want to be the strong woman and they want to engage with men who can express their feelings but still be a strong support.

Another stumbling block is the Bechdel test, a favorite of feminists to trot out to talk about movies (even though *The Lord of the Rings* somehow fails this test even with badasses like Éowyn, Arwen, and Galadriel). For those unfamiliar with the Bechdel test, it has three parts: 1) it must have more than one named female actor; 2) the female actors have to speak to each other; 3) the communication has to be about something other than men.[67] It is shocking to see that almost 43% of movies fail this test (bechdeltest.com, 2023), which unfortunately means that there are a whole lot of movies out there

---

[67] It's been a while since I've seen the *Lord of the Rings* movies, but I imagine the movies fail because the female characters never speak to one another. I don't think they're even in the same room together ever.

which use women as set pieces to move the story along without having any meaningful female relationships that don't revolve around men.

When looking at horror movies though, one finds that they have a seventy percent passing rate (N, 2023). The first part is pretty easy for horror movies to pass: if you have your Final Girl survivor, she's usually going to have a female best friend to be her foil, the anti-good girl. Sometimes there are more than just those two women, and all the better. Once the killer, or evil force starts to take out the characters, the women characters usually have a lot to talk about other than boys, like, staying alive or running away. This isn't to say that the Bechdel Test is the end all be all for whether or not horror movies are misogynistic. But it is another compelling argument in favor of horror movies.

If women are doing a lot more than just ogling men and talking to their girlfriends about it in horror movies, then are they dying more than men in horror movies? Surprisingly no, they are not. Men are much more likely to die in horror films, even slashers which are often interpreted as punishing women who step outside accepted gender norms and boundaries. The differences are in how the deaths are conducted (quick vs. a drawn-out chase scene), how memorable they are, and how long the camera lingers after the death.

Perhaps one of the reasons women seem to be consistently victimized or in danger in horror films is because that is the reality that most women face. From a young age, women are instructed to not walk home alone, especially not at night; to avoid strangers, especially men; to park under streetlights or near exits; to carry their keys between their knuckles or get pepper spray; and any number of other rules in order to stay safe from men. In horror films the women are on guard for, and running away from, male villains much like in real life. Women, feminists in particular, get so caught up in the fact that these women are being chased and tortured on screen that they forget that this is representative of the very real events and fears that women experience in the real world. Much like how Mystique being choked by Apocalypse should not have upset people because it is exactly what happens in superhero movies, and of what would happen in real life if a stronger villain met up with a weaker superhero. People should maybe not be so upset about women suffering in horror movies.

In Leigh Whannel's 2020 *Invisible Man*, Cecilia, played by Elisabeth Moss, attempts to navigate a world made scary by her abusive ex-husband. She is also blamed for the things that happen to her by her own sister. Cecilia has to exist in a world that may be very familiar to women who have struggled with exiting abusive relationships or having their traumatic stories denied by others. Cecilia eventually comes out triumphant through her own wits and strengths, but not without suffering. Notably Whannel, a man, did not write the character of Cecilia on his own. He used the input of other women to guide her actions and behaviors, and relied heavily on Elisabeth Moss to construct a believable female character who acts like a real person and not some hyper-masculinized woman that men can identify more with.

Whannel also noted that "The scarier monster is the one you can imagine in your own life. It's not a fanged beast, it's the guy next door." (Creahan, 2020). Taking into consideration the facts that show women are more likely to be abused, assaulted, or murdered by men they know in some capacity, this is truly real-life horror for women. Watching and enjoying this film can be a part of the cathartic experience that some suggest women get from watching rape-revenge films, or it can be a part of the safe-thrill idea, where women can feel "safely scared" watching a movie that includes elements of realism because there is a known endpoint.

Are there problems with the portrayal of women in horror movies? Sure. You can point to numerous examples in this book of the male gaze, problematic representation, and the unreasonable expectations placed on women. But that doesn't mean the genre as a whole is misogynistic.

Horror has been a vehicle for reflecting society's fears since *House of the Devil* first appeared in 1896. Whether it is racist fears of immigrants or people in the early 1900's, aliens and radioactive monsters in the 1940's to the 1970's, a generalized fear of our fellow man in the 1980's and 1990's, or the time of torture porn after 9/11 when nothing was sacred anymore because of the national trauma we had faced, horror can be seen to reflect the zeitgeist of the times. Women are frequently the people starring in these movies, fighting for their lives alongside more or less inept men who sometimes hinder the women more than they help them. In more recent films, these women are also becoming more adept at maintaining their autonomy and power throughout the film. They are less Scream Queens and more savage avengers. There are not many women now who would resemble Barbara or Wendy or Sally from the earlier decades of horror. These women are tackling their fears head on, and in some ways, maybe that should be a clarion call to the rest of us.

While the women of horror films are battling male villains who try to subjugate or victimize them through violence, gaslighting, disbelief, and torture, real life women are also facing these very real issues every day. Women who are not believed when they come forward about sexual violence; women who are more likely to be victimized by someone they know rather than a stranger; women who are told the "rules" for how to stay safe; women who are blamed for their trauma when they do not follow these "rules". Perhaps these things might have been accepted previously in a society locked in on subjugating women to keep them in line under the guise of keeping them safe.

But not so anymore. Real life women are striking back at the men, at the established patriarchy, and at all of the others (some women included) who would try to hold them back and tell them what a "proper" girl would do. They are breaking these rules the same way Sidney Prescott did nearly three decades ago in *Scream*.

If anything, horror has given women a safe place to feel scared; a place to see women being the badass without having to rely on a male co-star; a place where the woman character has evolved from the early days of the Scream Queen slasher to the take-charge women of today who sometimes have sex and sometimes drink and aren't always the ideal version of the Final Girl. I hope that Final Girls and horror movies continue to evolve and make changes while still keeping true to their roots of allowing women to have triumphant spaces all to themselves.

The horror genre and the Final Girl have been evolving over decades, changing shape and growing and shedding things that no longer serve them. The horror genre loves women, in the same way society should love women: putting them in power, believing and supporting them, and ultimately, letting them dictate how they will survive and who will help them get there.

Here's to the horror genre, my very favorite.

Acknowledgments

There are so many people that I must thank for the creation of this book. I guess the first people I should thank are my parents for instilling a love of reading in me and for always encouraging me to write, no matter what it was. Love you guys.

Thank you to my sister, Laura, and my cousin, Veronica, for being repeatedly used as a sounding board for writing ideas, the cover and book design, and for the non-stop talk about horror movies. I'm glad y'all were there.

Thank you to Kyle for patiently listening to me ramble about my book no matter how outlandish or existential I got about its creation or completion. Thank you for watching horror movies every day even when you would rather have watched a rom-com. I promise we can diversify our movie watching habits now that this is done. Thank you most of all for all the support you have given me on this endeavor.

The writing group at Valley Cottage Library was a great resource to me for deciding if my writing was accessible to casual horror enjoyers or if I was getting too technical. Thank you to all of you, especially Sean, for your constructive feedback.

To everyone who has ever talked for any amount of time with me about horror and movies, especially Angel. You are an irreplaceable source of knowledge for random things about obscure horror movies. All of my internet friends who responded to my countless polls, who took my survey, or who encouraged my research project in general, I appreciate all of you so much.

The risk in writing an acknowledgements page is that you are bound to miss someone who was a key part – big or small – in the creation of your book child. If that's you, please realize that it wasn't intentional. Please don't send any creepy puppets, puzzle boxes, or masked men with chainsaws my way. Thanks.

About the Author

If I had wanted to write about myself, I would have written a memoir, but alas, here is the "About the Author" page. Instead of me writing something, I asked my loved ones to provide one thing about me they thought you should know. This is the result.

She used to eat salt by the handful, licking it out of her palm.
She likes to chase cryptids.
She is a determined person who tries her best at everything she does.
She eats her cereal without milk in it.
She is attending mortuary school and prefers the dead to the living.
She is hard on the outside but is really a big softie. (Thanks Mom).

Find me on
Instagram @modernmonstress
TikTok @modernmonstress
https://kimberlypinzon.wixsite.com/my-site-5

Appendices

Appendix A

Comments on the article written by Krystie Lee Yandoli

giving a synopsis that includes ALL major plot points as well as the ending? The last four paragraphs could be cut without any meaningful change to the point of her piece.

Like · Reply · 1 · 3w

Jared 'Sage' Juetten

With the whole Amber Heard/Johnny Depp thing going on right now I feel like "believe women" might be a tough theme to shove down people's throats...especially when the female protagonist straight murders her husband at the end even after she knows he wasn't the invisible Man. Justice? How about Just-leave-him?

Thanks for writing the WHOLE PLOT for me so I don't need to see this movie.

Like · Reply · 5 · 3w

Nick Schwaller

Well on the bright side, I won't check it out at all. Believe women, lol, yeah, sure. How about no and wait until proof of something, anything, comes in. And yeah, the whole Amber heard is the perfect example of what happens when accusations are enough to convict, no due process. What a joke

Like · Reply · 3 · 3w

Kaylee .M

I get paid over $90 per hour working from home with 2 kids at home. I never thought I'd be able to do it but my best friend earns over 10k a month doing this and she convinced me to try. The potential with this is endless. Heres what I've been doing, <(")

Copy Here.........>> Www.workbaar.Com♭

Please Remove (♭) when copy url

Like · Reply · 3w

The gaslighting of women is a huge problem, I agree. But if someone (man or woman) approached me telling me that their dead husband was alive, invisible, and wreaking havoc, I would "sure, Jan" the shit right out of them. Can't fault the other characters for that 🙃

Like · Reply · 6 · 3w

Nickolas Holloway

Yet another political smear campaign of a good movie. Movie critics have lost all credibility nowadays, and should be viewed as nothing more than whining crybabies. Not everything has to be political, just a good movie. Doesn't help that the whole #MeToo thing is just a witch hunt now. Unless there is proof then they shouldn't be believed because of how harmful it is to be wrongly "MeToo" someone.

Like · Reply · 3w · Edited

Trivitt Adams

LMAO! There are spoilers and then there are explain the entire movie! Wasn't gonig to go see the movie anyway but damn son, you ruined it for a lot of people that were.

Like · Reply · 1 · 3w

James Baxter

I'm not sure who is the more stupid; the morons complaining that you gave away the plot after saying there were spoilers or your commentary, which is some of the worst, factually incorrect, woke filled dross I've seen lately, and that's saying something in these days! How the hell you were given a job as a journalist is beyond me...

Like · Reply · 3w

James Cork

How "The Invisible Man" Shows The Horror Of Not Believing Women In The #MeToo Era

Yes... if your ex learns how to become invisible and mess with you, this movie will be a perfect lesson about belielving women who find themselves in that unique fictional scenario....

Like · Reply · 3 · 3w

Appendix B

Survey Questions

I conducted a survey of 78 participants who were found on Reddit, Instagram, and Facebook. Most questions were multiple choice options, and some had the option to write an "Other" answer if the participant chose. The questions included in the survey were:

1. What is your gender?
2. What is your age?
3. What was the first horror movie that you saw?
4. What is your favorite horror movie?
5. How much do you like horror movies?
6. What horror subgenres do you recognize?
7. Which horror subgenres have you watched?
8. Do horror movies frighten you?
9. Do you feel that horror movies have misogynistic tendencies?
10. Are women more likely than men to get killed in horror movies?
11. Are women more likely than men to be shown naked in horror movies?
12. Is a woman being naked more likely to be sexual than a man being naked?
13. Final Girls are defined as "last girl(s) or woman alive to confront the killer, ostensibly the one left to tell the story." They are sometimes the only person left alive by the end of the film. Are Final Girls survivors or victims?

14. Are women in horror movies portrayed differently than in other movies?

15. If you answered yes, how are they portrayed differently?

16. If you are a male, do you feel like you are able to identify with a female protagonist in a horror movie?

17. Rape-revenge horror movies are typically framed as a "Triumph" for women because they ultimately destroy the men who assaulted them. Are they actually triumphing, or is this an over reliance on the "broken woman trope" which alleges that women are weak until after they are traumatized?

18. Are horror movies more or less satisfying when the protagonist goes through pain, suffering, and loss?

19. Do you notice social issues commentary embedded in horror movies?

20. Are horror movies made for men or women?

https://s.surveyplanet.com/o2h7mr40

Appendix C

DeadMeat YouTube Channel Death Tally Counts

Friday 13th 7	Revenge	Friday 13th Part 3 1982	Friday 13th Part 2 1981	Friday 13th 1980		
7F/17M	3M	5W/7M	4W/5M	5W/5M		

Midsommar	Day of the Dead 1985	13 Ghosts	Wrong Turn 2021	Babysitter Killer Queen		
4W/8M	10M	2W/10M	2W/13M	8W/10M		

House of 1000 Corpses	Rec 2	Wrong Turn 3	Wrong Turn	I Still Know What You Did Last Summer		
7W/7M	13M	2W/13M	3W/6M	2W/8M		

I Know What You Did Last Summer	Tourist Trap	Sinister 2	Sinister	Quarantine	Rec	
2W/4M	3W/3M	10W/11M	10W/10M	9W/14M	6W/10M	

The Cabin in the Woods	Ready or Not	Slumber Party Massacre	Candyman 1992	Final Destination 5		
8W/42M	7W/8M	6W/6M	3W/3M	2W/6M		

Final Destination	The Descent	Jaws Revenge	Jaws 3D	Jaws 2	Jaws	
2W/3M	5W/1M	1W/1M	5M	3W/4M	1W/4M	

Sleepaway Camp	TCM Beginning	TCM 2003	TCM 1974	My Bloody Valentine		
3W/4M	4W/7M	2W/5M	1W/4M	10W/18M		

Halloween 2017	Silent Night 2012	Silent Night Deadly Night 1984	Evil Dead 2013	Evil Dead 1981		
4W/14M	6W/11M	4W/9M	5W/3M	6W/2M		

Halloween 2	Halloween	Unfriended	Saw 6	Saw 4	Saw 3	Saw
5W/5M	3W/2M	3W/3M	2W/8M	2W/8M	3W/6M	6M

Saw 2	Saw 5	Nightmare on Elm St 2010	NES 4 1988	NES 2 1985	NES 1984	
2W/7M	4W/9M	2W/3M	3W/3M	3W/7M	2W/2M	

Black Christmas 2006	Black Christmas 1974	Alien	Trick r Treat	Scream 6	Meg	
11W/7M	5W/2M	1W/4M	4W/16M	3W/10M	2W/18M	

Winnie the Pooh	Urban Legend Bloody Mary	Urban Legend Final Cut	Urban Legend Kill Count	Critters Attack		
8W/3M	2W/5M	3W/4M	3W/7M	1W/10M		

The Collection	the Collector	Scream 2022	Scream 4	Scream 3
43W/32M	3W/5M	3W/5M	9W/5M	4W/6M

Scream 2	Scream	Malignant	Child's Play	
4W/6M	2W/5M	24W/22M	1W/5M	

WolfCreek	Would You Rather	Halloween kills	Slither	Tremors
3W/2M	2W/7M	5W/7M	37W/52M	1W/7M

Candyman 2022	Hellraiser	Friday 13th reboot	Freddy vs. Jason	
8W/9M	1W/6M	5W/10M	7P/21M	

Spiral	Jason X	Jason Goes to Hell	
1W/7M	7W/21M	7W/16M	

Works Cited

Introduction
Category: American rape and revenge films. Wikipedia. (7 May 2021).
https://en.wikipedia.org/wiki/Category:American_rape_and_revenge_films
Hall, Gladys. (January 1931). The feminine love of horror. Motion Picture Classic.
https://beladraculalugosi.wordpress.com/2011/09/28/the-feminine-love-of-horror/
Krishnan, Vidya. (2023, June 2nd). *In India's gang rape culture, all women are victims.* The New York Times.
https://www.nytimes.com/2023/06/02/opinion/india-women-rape.html#:~:text=In%202011%20a%20woman%20was,rapes%20were%20reported%20to%20authorities.

Pathak, S. & Frayer, L. (2019, December 29th). *What headlines and protests get wrong about rape in India.* NPR.
https://www.npr.org/sections/goatsandsoda/2019/12/29/791734411/what-headlines-and-protests-get-wrong-about-rape-in-india

The Tangents: Is the shark in *Jaws* male or female?
Teague, Savanna. (Accessed 10/30/2022) https://the-take.com/read/why-does-bruce-the-sharkas-gender-matter-in-jaws#:~:text=Jaws%20does%20not%20label%20its,presents%20the%20shark%20as%20male.

What is Horror Anyway?
BBC News. (February 27th, 2006). Slovakia angered by horror film. http://news.bbc.co.uk/1/hi/entertainment/4754744.stm
Flowers, Maisy. (June 20th, 2020). Night of the Living Dead: why George Romero rewrote Ben's character.
https://screenrant.com/night-living-dead-george-romero-rewrite-ben-character-duane-jones-reason/
Keetley, Dawn. (January 30th, 2016). Everest and Frozen: exploring the edges of horror.

http://www.horrorhomeroom.com/everest-and-frozen-exploring-the-edges-of-horror/

Marshall, Rick. Obscenity Case Files: *Jacobellis v. Ohio* ("I know it when I see it") https://cbldf.org/about-us/case-files/obscenity-case-files/obscenity-case-files-jacobellis-v-ohio-i-know-it-when-i-see-it/#:~:text=In%20his%20concurring%20opinion%20in,I%20see%20it.%E2%80%9D%20statement.

Wilkinson, Alissa. (July 22nd, 2017). George Romero didn't mean to tackle race in Night of the Living Dead, but he did anyway. https://www.vox.com/culture/2017/7/22/15985492/night-of-living-dead-movie-week-george-romero-zombies-get-out-jordan-peele

The Tangents: What are Your Odds of Dying in a Hostel?

Stimac, Blake. (January 27th, 2020). Eli Roth's Dangerous Inspiration for *Hostel* Explained. https://screenrant.com/eli-roth-hostel-movie-murder-tourism-inspiration-explained/#:~:text=In%20an%20interview%2C%20Eli%20Roth,%2C%20advertising%20%22murder%20vacations.%22

National Safety Council. (2021). Odds of Dying. https://injuryfacts.nsc.org/all-injuries/preventable-death-overview/odds-of-dying/

McLaughlin, Kelly. (2023, May 26th). The chance of getting bitten by a shark while you're swimming at the beach are surprisingly low. Insider. https://www.insider.com/shark-attacks-what-are-odds-of-getting-bitten-2018-7

Nordentoft, Merete and Wandall-Holm, Nina. (July 12th, 2003). 10 year follow up study or mortality among users of hostels for homeless people in Copenhagen. https://www.ncbi.nlm.nih.gov/pmc/articles/PMC164916/

A (Brief) History of Horror

https://ncttheatre.com/blog/women-theatre-historical-look

Antrobus, Helen; Tidd, Natasha; and Westrop, Sara. (October 21st, 2017). The 4 forgotten women who built horror. https://fyeahhistory.wordpress.com/2017/10/21/the-4-forgotten-women-that-built-horror/

Bose, Swapnil Dhruv. (July 17th, 2021). *Alice Guy-Blaché: the enormous legacy of the first-ever female filmmaker*. Far Out

Magazine. https://faroutmagazine.co.uk/alice-guy-blache-the-enormous-legacy-of-the-first-ever-female-filmmaker/

Bose. Swapnil Dhruv. (December 2nd, 2021). *The pioneering women who helped build Hollywood.* Far Out Magazine. https://faroutmagazine.co.uk/the-women-who-helped-build-hollywood/

dimofhorror. (September 6th, 2016). Every found footage movie. IMDb. https://www.imdb.com/list/ls063914804/?sort=release_date,asc&st_dt=&mode=detail&page=1

James, Caryn. (March 20th, 2019*). Lois Weber: the trailblazing director who shocked the world.* BBC. https://www.bbc.com/culture/article/20190318-lois-weber-the-trailblazing-director-who-shocked-the-world

Keetley, Dawn. (August 12th, 2016) Thirteen Women (1932): the slasher that started it all. http://www.horrorhomeroom.com/thirteen-women/

Lauzen, Martha M. (2017). It's a man's (celluloid) world: portrayals of female characters in the top 100 films of 2016. *Center for the Study of Women in Television and Film.* https://womenintvfilm.sdsu.edu/wp-content/uploads/2017/02/2016-Its-a-Mans-Celluloid-World-Report.pdf

Lauzen, Martha M. (2021). The celluloid ceiling: behind-the-scenes employment of women on the top U.S. films of 2020. *Center for the Study of Women in Television and Film.* https://www.nywift.org/wp-content/uploads/2021/11/2020-Celluloid-Ceiling-Report.pdf

Lauzen, Martha. M. (2022). The celluloid ceiling in a pandemic year: employment of women on the top U.S. films of 2021. *Center for the Study of Women in Television and Film.* https://womenintvfilm.sdsu.edu/wp-content/uploads/2022/01/2021-Celluloid-Ceiling-Report.pdf

Morris, Amanda. (April 1st, 2020). Golden age of Hollywood was not so golden for women. *Northwestern Now.* https://news.northwestern.edu/stories/2020/03/golden-age-of-hollywood-was-not-so-golden-for-women/

New York Film Academy, The. (July, 21st2022). https://www.nyfa.edu/student-resources/how-horror-movies-have-changed-since-their-beginning/#:~:text=Just%20a%20few%20years%20after,be%20the%20first%20horror%20movie.

Solomon, Josh. (2020). A brief history of early horror. https://viterbi-web.usc.edu/~jdsolomo/itp104/assignment_06/home.html

Thirteen Women. IMDB. https://www.imdb.com/title/tt0023582/?ref_=nv_sr_srsg_0

Vatsal, Radha. (March 29th, 2016). The forgotten female actions star of the 1910s. *The Atlantic.* https://www.theatlantic.com/entertainment/archive/2016/03/the-forgotten-female-action-stars-of-the-1910s/475635/

Weitzman, Elizabeth. (April 26th, 2019). *A century late, a giant of early cinema gets her closeup.* The New York Times. https://www.nytimes.com/2019/04/26/movies/alice-guy-blache-be-natural.html

The Tangents: Why are people so high on "elevated horror"?

Bradley, Laura. (December 17th, 2019). This was the decade horror got "elevated". https://www.vanityfair.com/hollywood/2019/12/rise-of-elevated-horror-decade-2010s

Cotroneo, Vincent. (February 15th, 2022). These are some of the elevated horror films which helped make horror so hot. https://movieweb.com/elevated-horror-films/

Gislason, Lor. (September 4th, 2021). Why elevated horror is an unnecessary and elitist term. https://horrorobsessive.com/2021/09/04/why-elevated-horror-is-an-unecessary-and-elitist-term

Mendelsohn, Jon. (June 10th, 2020). Elevated horror: the movies of a subgenre that may not exist. https://www.cbr.com/elevated-horror-subgenre-examples/

Lattanzio, Ryan. (October 12th, 2022). John Carpenter has no idea what 'elevated horror' means. https://www.indiewire.com/2022/10/john-carpenter-elevated-horror-1234771814

Sharf, Zack. (October 31st, 2022). Jordan Peele says he's not 'trying to make elevated' genre movies: 'that's a trap I don't quite appreciate'. https://variety.com/2022/film/news/jordan-peele-rejects-elevated-horror-label-trap-12345418851

Zigler, Brianna. (May 31st, 2022). *Men* and the end of elevated horror. https://www.pastemagazine.com/movies/men-elevated-horror-alex-garland/

Who is the Final Girl?

CBS. (May 10th, 2018). How Abby Sciuto defied convention and changed TV.
https://www.cbs.com/shows/ncis/news/1008514/how-abby-sciuto-defied-convention-and-changed-tv/

Geena Davis Institute on Gender in Media. (Accessed December 21st, 2022). Female characters in film and TV motivate women to be more ambitious, more successful, and have even given them the courage to break out of abusive relationships.
https://seejane.org/gender-in-media-news-release/female-characters-film-tv-motivate-women-ambitious-successful-even-given-courage-break-abusive-relationships-release/

Geena Davis Institute on Gender in Media. (Accessed December 21st, 2022). The Scully effect: I want to believe…in STEM. https://seejane.org/wp-content/uploads/x-files-scully-effect-report-geena-davis-institute.pdf

Heldman, Caroline. (2016). Hitting the bullseye: reel girl archers inspire real girl archers. https://seejane.org/wp-content/uploads/hitting-the-bullseye-reel-girl-archers-inspire-real-girl-archers-full.pdf

Kiste, Gwendolyn. (January 6th, 2022). In defense of Wendy, Barbara, and the traumatized women of horror cinema. Nightfire.

Reilly, Kaitlin. (December 16th, 2016). Why this *Scream* character was so good for women. https://www.refinery29.com/en-us/2016/12/130121/sidney-prescott-scream-movies-heroine

Wijaszka, Zofia. (October 8th, 2019). A final girl trope in horror films: then and now.
https://intheirownleague.com/2019/10/08/a-final-girl-trope-in-horror-films-then-and-now/

Young, Cate. (March 6th, 2020). What happened to cinema's virginal final girl?
https://editorial.rottentomatoes.com/article/what-happened-to-cinemas-virginal-final-girl/

The Tangents: Do final girl characteristics mimic the "Ideal Victim" characteristics looked for in law enforcement?

Gramlich, John. (November 20th, 2020). What the data says (and doesn't say) about crime in the United States.

https://www.pewresearch.org/fact-tank/2020/11/20/facts-about-crime-in-the-u-s/

National Coalition Against Domestic Violence. https://ncadv.org/STATISTICS

Schwobel-Patel, Christine. (August 5th, 2015). Nils Christie's "ideal victim" applied: from lions to swarms. https://criticallegalthinking.com/2015/08/05/nils-christies-ideal-victim-applied-from-lions-to-swarms/

Uniform Crime Report. 2012. Arrests by sex, 2012. https://ucr.fbi.gov/crime-in-the-u.s/2012/crime-in-the-u.s.-2012/tables/42tabledatadecoverviewpdf/table_42_arrests_by_sex_2012.xls

When You Gaze at the Man the Man Gazes Back at You…or Something

Baroni, Lorenzo. (September 11th, 2022). Why does Ash malfunction in Alien. ScreenRant. https://screenrant.com/why-does-ash-malfunction-in-alien/

Bertram, Colin. (June 18th, 2019). How women in horror movies keep us coming back for more. https://www.biography.com/news/women-in-horror-movies

Billson, Anne. (October 25th, 2018). When did you last see a man begging for his life in a horror movie? The Guardian. https://theguardian.com/film/2018/oct25/when-did-you-last-see-a-man-begging-for-his-life-in-a-horror-movie

Evans, Alan. (October 31st, 2016). *Tippi Hedren: Alfred Hitchcock sexually assaulted me.* The Guardian. https://www.theguardian.com/film/2016/oct/31/tippi-hedren-alfred-hitchcock-sexually-assaulted-me

Lemon, Grace. Female victimization in the horror genre. https://www.sceneandheardnu.com/female-victimization-in-horror

Lord, Annie. (March 10th, 2020). 'Torture the women!': how horror's final girls are turning the tables on misogyny. https://www.independent.co.uk/arts-entertainment/films/features/invisible-man-horror-movie-midsommar-witch-final-girls-hitchcock-a9390356.html

Male Gaze. Wikipedia. https://en.wikipedia.org/wiki/Male_gaze#cite_note-Stack-13

Media Studies. (Accessed November 7th, 2022). The male gaze. https://media-studies.com/male-gaze/

Molitor, F., & Sapolsky, B. S. (1993). Sex, violence, and victimization in slasher films. *Journal of Broadcasting & Electronic Media, 37*(2), 233–242. https://doi.org/10.1080/08838159309364218

Oxford Reference. Male gaze. https://www.oxfordreference.com/view/10.1093/oi/authority.201108 03100128610

Sur, Debadrita. (October 5th, 2021). Is the representation of women in horror gradually changing? https://faroutmagazine.co.uk/is-the-representation-of-women-in-horror-gradually-changing

Vanbuskirk, Sarah. (October 17th, 2022). What is the male gaze? https://www.verywellmind.com/what-is-the-male-gaze-5118422

White, Edward. (April 26th, 2021). *The dark side of an auteur: on Alfred Hitchcock's treatment of women.* Literary Hub. https://lithub.com/the-dark-side-of-an-auteur-on-alfred-hitchcocks-treatment-of-women/

The Tangents: The Case for Paris Hilton

Collet-Serra, J. (Director). (2022). *House of Wax.* [2022 Blu-ray release from SHOUT! Factory]. Village Roadshow Pictures, Dark Castle Entertainment.

Gaslight the Women

Chavez Muiron, Regina. (February 28th, 2020). Women in horror: fighting for your life and voice. https://aiptcomics.com/2020/02/28/women-in-horror-fighting-for-your-life-and-voice/

Jacobs, Matthew. (March 2nd, 2020). Why does no one in horror movies believe the female protagonist? https://www.huffpost.com/entry/the-invisible-man-horror-trope-female-protagonist

Leotta, Allison. (2018, October 3rd). *I was a sex-crimes prosecutor. Here's why 'he said, she said' is a myth.* Time Magazine. https://time.com/5413814/he-said-she-said-kavanaugh-ford-mitchell/

Searles, Jourdain. (March 11th, 2020). What *The Invisible Man* gets right about women in horror. https://www.gq.com/story/what-the-invisible-man-gets-right-about-women-in-horror

Yandoli, Krystie Lee. (February 27th, 2020). How "The Invisible Man" shows the horror of not believing women in the #metoo era.
https://www.buzzfeednews.com/article/krystieyandoli/the-invisible-man-me-too

Is That a Knife or Are You Just Happy to See Me?

Barone, Matt. (2013, October 31st). *Fact check: Do black characters always die first in horror movies?* Complex.
https://www.complex.com/pop-culture/a/matt-barone/black-characters-horror-movies

Flowers, Maisy. (June 20th, 2020). Night of the Living Dead: Why George Romero rewrote Ben's character. ScreenRant.
https://screenrant.com/night-living-dead-george-romero-rewrite-ben-character-duane-jones-reason/

Glosserman, Scott. (Director). (2006). *Behind the mask: The rise of Leslie Vernon.* [Film]. Anchor Bay Entertainment.

Kane, Joe. (August 31st, 2010). How casting a black actor changed 'Night of the Living Dead'. The Wrap.
https://www.thewrap.com/night-living-dead-casting-cult-classic-20545/

Subisatti, A. & West, A. (Hosts). (October 31st, 2017). Undead Walking: Night of the Living Dead (1968), Dawn of the Dead (1978) and Day of the Dead (1985). (Episode 54). [Audio Podcast Episode]. In *The Faculty of Horror.* Produced independently.

Wilkinson, Alissa. (July 22nd, 2017). George Romero didn't mean to tackle race in Night of the Living Dead, but he did anyway.
https://www.vox.com/culture/2017/7/22/15985492/night-of-living-dead-movie-week-george-romero-zombies-get-out-jordan-peele

The Tangents: Show Me Your O (Shit) Face

Avery, Dan. (April 5th, 2021). Most people cannot tell the difference between screams of joy and screams of fear because they both have similar acoustic features, study finds.
https://www.dailymail.co.uk/sciencetech/article-9437799/Most-people-tell-difference-screams-joy-fear-sound-similar.html

Berman, Robby. (April 6th, 2021). Is it happiness or fear? Why some screams may confuse us.
https://www.medicalnewstoday.com/articles/is-it-happiness-or-fear-why-some-screams-may-confuse-us

What About the Female Gaze?
Bergerson, Samantha. (April 27th, 2023). IndieWire. *Sigourney Weaver is done with 'Alien' franchise: "I put in my time in space."* https://www.indiewire.com/features/general/sigourney-weaver-done-alien-franchise-1234832954/

Billson, Anne. (August 2nd, 2019). The Guardian. *Does the 'female gaze' make sexual violence on film any less repugnant?* https://theguardian.com/film.2019/august.02/the-female-gaze-dies-it-make-sexual-violence-on0film-any-less-repugnant

Connors, Graham. (October 29th, 2016). Head Stuff. *Women in danger: dissecting the voyeuristic gaze of the modern slasher film.* https://headstuff.org/entertainment/film/women-in-danger-dissecting-the-voyeuristic-gaze-of-the-modern-slasher/

Forster, Stefani. (June 12th, 2018). Medium. *Yes, there's such a thing as a 'female gaze'. But it's not what you think.* https://medium.com/truly-social/yes-theres-such-a-thing-as-the-female-gaze-but-it-s-not-what-you-think-d27be6fc2fed

Gazvoda, Melanie Rose. (April 2nd, 2023). By Arcadia. *Horror film 101: surviving the male gaze.* https://www.byarcadia.org/post/horror-film-101-surviving-the-male-gaze

Helford, Elyce Rae. (Accessed July 12th, 2023). The Take. *Is a female gaze possible in cinema?* https://the-take.com/read/is-a-female-gaze-possible-in-cinema

Kavanagh, Emily. (October 9th, 2022). *How 'Jennifer's Body' unsettles the way we typically view female characters.* Collider. https://collider.com/jennifers-body-female-gaze-horror-explained/

Lemon, Grace. (Accessed July 12th, 2023). Scene and Heard. *Female victimization in the horror genre.* https://www.sceneandheardnu.com/female-victimization-in-horror

Mead, Eleanor. (October 13th, 2022). *What is the female gaze in film?* Video Librarian. https://videolibrarian.com/articles/essays/the-meaning-of-female-gaze-in-film/

Schneider, Maggie. (March 16th, 2022). The Science Survey. *See you seeing me: the female gaze in cinema.* https://thesciencesurvey.com/arts-entertainment/2022/03/16/seeing-you-seeing-me-the-female-gaze-in-cinema/

221

Sims, David. (August 2nd, 2018). The Atlantic. *The value of the 'female gaze' in film.*
https://www.theatlantic.com/entertainment/archive/2018/08/female-gaze-lincoln-center-series-women-cinematographers/566612/

Smith, Gwendolyn. (February 22nd, 2020). The Guardian. *Like a natural woman: how the female gaze is finally bringing real life to the screen.*
https://www.theguardian.com/culture/2020/feb/22/the-female-gaze-brings-a-welcome-touch-of-reality-to-art

Sur, Debadrita. (October 5th, 2021). Far Out Magazine. *Is the representation of women in horror changing gradually?*
https://faroutmagazine.co.uk/is-the-representation-of-women-in-horror-changing-gradually/

Telfer, Tori. (August 2nd, 2018). Vulture. *How do we define the female gaze in 2018?*
https://www.vulture.com/2018/08/how-do-we-define-the-female-gaze-in-2018.html

Vosper, Amy Jane. (July 2014). *Film, fear, and the female: an empirical study of the female horror fan.* Off Screen.
https://offscreen.com/view/film-fear-and-the-female

Zukowski, Kelsey. (September 3rd, 2021). *How the representation of female killers in horror showcases the struggle and strengths of womanhood.* Collider. https://collider.com/female-killers-in-horror-movies-representation/

The Tangents: Is it Illegal to Show a Penis on Television?
Bryant, Kenzie. (March 17th, 2022). *The penile code: what TV's increase in full-frontal male nudity really means.* Vanity Fair.
https://www.vanityfair.com/hollywood/2022/03/minx-euphoria-tvs-increase-in-full-frontal-male-nudity-prosthetics

Kato, Brooke. (February 7th, 2022). *The power of the dong: the year the penis was unleashed in Hollywood.* NY Post.
https://nypost.com/2022/02/07/power-of-the-dong-hollywood-unleashed-the-penis-this-year/

Taylor, Magdalene. (April 2022). *Show us your real dicks, you movie star cowards.* Mel Magazine.
https://melmagazine.com/en-us/story/golden-age-of-male-nudity-fraud

Finding Catharsis in Strange Places

Austin, Isobella. (February 18th, 2021). *'Rape-revenge' films are changing: they now focus on women, instead of their dads.* The Conversation. https://theconversation.com/rape-revenge-films-are-changing-they-now-focus-on-the-women-instead-of-their-dads-155456

Leite, Natalia. (November 8th, 2017). *Why women need to tell rape stories.* Talkhouse. https://www.talkhouse.com/why-women-need-tell-rape-stories/

Machado, Carmen Maria. (January 29, 2021). The New Yorker. https://www.newyorker.com/culture/cultural-comment/how-promising-young-woman-refigures-the-rape-revenge-movie

McAndrews, Mary Beth. (October 8th, 2019). *[Through her eyes] the history of rape-revenge films and the importance of female directors.* Bloody Disgusting. https://bloody-disgusting.com/editorials/3586210/eyes-history-rape-revenge-films-importance-female-directors/

McAndrews, Mary Beth. *When the party's over: how women reclaim the rape-revenge story.* Girls on Tops. https://www.girlsontopstees.com/read-me/2019/10/25/when-the-partys-over-how-women-reclaim-the-rape-revenge-story

Nugent, Annabel. (April 27th, 2021). *'I wanted to use violence in excess': how women filmmakers reclaimed the revenge movie.* Independent. https://www.independent.co.uk/arts-entertainment/films/features/revenge-films-promising-young-woman-b1836483.html

Subissati, Andrea. (January 10th, 2018). *Exclusive interview: the women behind "M.F.A." add a female voice to rape-revenge.* Rue Morgue. https://rue-morgue.com/exclusive-interview-the-women-behind-m-f-a-add-a-female-voice-to-rape-revenge/

Wilson, Lena. (May 11th, 2018). Revenge *tries to elevate the rape-revenge movies, but is the genre worth saving?* Slate. https://slate.com/culture/2018/05/revenge-and-the-case-against-rape-revenge-films.html

Wilson, Lena. (January 14th, 2021). Rape-revenge tales: cathartic? Maybe. Incomplete? Definitely. https://www.nytimes.com/2021/01/14/movies/rape-revenge-films-flaws.html

The Tangents: What's Your Favorite Scary Movie?

ElmStreet1985. (2021). Why do girls love the SCREAM movies so much? [Online forum post]. Reddit
https://www.reddit.com/r/horror/comments/s5g589/why_do_girls_love_the_scream_movies_so_much/

Why Do We Think Women Die More in Horror Than Men?

@DeadMeat. (n.d). Dead Meat. Youtube. Retrieved from https://www.youtube.com/@DeadMeat

Cowan, Gloria & O'Brien, Margaret. (1990). Gender and survival vs. death in slasher films: A content analysis. Sex Roles. 23. 187-196. 10.1007/BF00289865.

Duca, Lauren. (2013, October 13). White women are victimized more than any other demographic in horror movies. HuffPost. https://www.huffpost.com/entry/horror-film-deaths_n_4178359

The Tangents: But What Does the Internet Think?

Username deleted. (2016). Why do more men get tortured and killed in movies than women? (not being sexist, just read the text). [Online forum post]. Reddit.
https://www.reddit.com/r/movies/comments/4xl8a9/why_do_more_men_get_tortured_and_killed_in_movies/?rdt=35944

goreboy. (2012). Why do women survive horror films, but men don't? [Online forum post]. Reddit.
https://www.reddit.com/r/horror/comments/1c9rm5/why_do_women_survive_horror_films_but_men_dont/

kently7. (2015). Why do most horror films have a female protagonist? [Online forum post]. Reddit
https://www.reddit.com/r/horror/comments/4bz53a/why_do_most_horror_films_have_a_female_protagonist/

BBC. (n.d.). Horror, from the point of view of the female gaze. Retrieved from
https://www.youtube.com/watch?v=DWow4o3V8IA

Sad Girls Don't Die, They Become Villains

Chun, Rene. (January 6th, 2020). *Female fugitives: why is 'pink-collar' crime on the rise?* The Guardian.
https://www.theguardian.com/us-news/2020/jan/06/female-fugitives-women-crime-rates-rise

Miller, Alyssa. (August 9th, 2021). Girlhood to monstress: how women in horror have always been the hero and the villain. https://nofilmschool.com/women-as-hero-and-villain-in-horror

Paxton, Kelly. https://pinkcollarcrime.com/

Salzman, Eva. (November 17th, 2021). Horror genre has relied on female sexuality. The Ithacan. https://theithacan.org/42294/life-culture/popped-culture/lc-horror-genre-has-relied-on-female-sexuality/#:~:text=In%20the%20Molitor%20and%20Sapolsky,die%20on%20screen%20than%20men.

The Tangents: *The Descent* (2005) is a Feminist Horror Masterpiece

Erbland, Kate. (March 17th, 2020). *Stream of the day:* 'The Descent' *is not afraid to give us imperfect women.* Yahoo Entertainment. https://www.yahoo.com/entertainment/stream-day-descent-now-amazon-130021164.html?guccounter=1&guce_referrer=aHR0cHM6Ly93d3cuZ29vZ2xlLmNvbS8&guce_referrer_sig=AQAAAKSaFckeU4Fqgq754kfMzmCA2kOCxESKw-Ihq5TnOo6sysx9NxoaVpr8pemhbbybx9Wl73JrB06kBi9GVncKoD7H0lij8bw5322Wxw9Ke-mshnfcTO3zGdfH4xbaMbUAKEIJzP3fR1ACTC0dT1cNhdzuN6paOcvoDRaQkyNElTa0

Handler, Rachel. (February 24th, 2021). How *The Descent* ended up with two famously bleak endings. The Vultre. https://www.vulture.com/2021/02/the-descent-alternate-ending-explained-by-neil-marshall.html

Mikulec, Sven. (May 24th, 2022). Neil Mrshall's '*The Descent*': humans are the scariest things. Cinephilia & Beyond. https://cinephiliabeyond.org/descent/

Why do Women Love to Scream?

Anxiety & Depression Association of America. (Accessed December 5th, 2022). Women and Anxiety. https://adaa.org/find-help-for/women/anxiety

Backus, Fred. (October 31st, 2021). CBS News poll: one in two Americans enjoy watching scary movies. https://www.cbsnews.com/news/scary-movies-opinion-poll/

Battaglio, Stephen. (January 5th, 2016). *Investigation Discovery becomes top cable channel for women with true crime all*

the time. Los Angeles Times.
https://www.latimes.com/entertainment/envelope/cotown/la-et-ct-investigation-discovery-20160105-story.html

Civic Science. (October 26th, 2017). 13 insights about horror movie fans. https://civicscience.com/horror-movie-fan-facts/

Clancy, Michelle. (August 6th, 2019). *Discovery tops among female viewers in US*. RapidTVNews.
https://www.rapidtvnews.com/2019080656886/discovery-tops-among-female-viewers-in-us.html#axzz7oy5hzVdm

Griffiths, Mark. (October 29th, 2015). Why do we like watching scary films?
https://www.psychologytoday.com/us/blog/in-excess/201510/why-do-we-watching-scary-films

Haidt, J., McCauley, C., & Rozin, P. (1994). Individual differences in sensitivity to disgust: A scale sampling seven domains of disgust elicitors. Personality and Individual Differences, 16, 701-713.

Klein, J. (2019, August 27th). Oxygen's true crime rebrand keeps paying off. Fortune. https://fortune.com/2019/08/27/oxygen-true-crime-rebrand-two-years-later/

National Crime Victimization Survey. (September 2010). United States Department of Justice.
https://bjs.ojp.gov/content/pub/ascii/vdhb.txt

Spines, Christine. (July 24th, 2009). Chicks dig scary movies. Entertainment Weekly.
https://ew.com/article/2009/07/24/chicks-dig-scary-movies

Pew Research Center. Accessed 11/23/2022.
https://www.pewresearch.org/religion/religious-landscape-study/

Rubin, Rebecca. (October 25th, 2018). Diverse audiences are driving the horror box office boom.
https://variety.com/2018/film/box-office/horror-movies-study-1202994407/

Turvey, Malcolm. (October 30th, 2018). Why do we like horror movies? https://now.tufts.edu/print/articles/why-do-we-horror-movies

Unified Crime Report. Accessed November 23rd, 2022. FBI. https://ucr.fbi.gov/crime-in-the-u.

United States Census Bureau, 2020. (Accessed December 5th, 2022).
https://www.census.gov/quickfacts/fact/table/US/LFE046220

Wallace, Lindsay Lee. (July 18th, 2021). *Horror media has become a vital and transgressive artistic inflection point for women.* Blood Knife. https://bloodknife.com/why-women-watch-horror/

Is Horror Really Misogynistic Trash?

Bechdel Test Movie List. (Accessed September 10th, 2023). https://bechdeltest.com/

Burgos, Danielle. (2019, July 29th). 46 horror movies that pass the Bechdel Test, proving that this genre's women have agency. Bustle. https://www.bustle.com/p/46-horror-movies-that-pass-the-bechdel-test-proving-that-this-genres-women-have-agency-18194512

Cowan, Gloria & O'Brien, Margaret. (August 1990). Gender and survival vs. death in slasher films: a content analysis. *Sex Roles, Vol. 23, Nos. 3/4, 1990,* 187-196.

Creahan, Danica. (March 16th, 2020). The "Invisible Man" makes women visible again in horror films. https://asiamedia.lmu.edu/2020/03/16/the-invisible-man-makes-women-visible-again-in-horror-films/

Friedman, L., Daniels, M., & Blinderman, I. (Accessed August 29th, 2023). Hollywood's gender divide and its effect on film. Pudding. https://pudding.cool/2017/03/bechdel/#:~:text=On%20bechdeltest.com%2C%20a%20site,%2C%20write%20what%20you%20know).

Gooden, Tai. (March 31st, 2022). Women in slasher films: their deaths, evolution, and potential future.

N. Erika. (January 2023). Data analysis of female representation in movies – has the # of Bechdel Test-passing movies increased? Medium. https://medium.com/@nagainagai.e/data-analysis-of-female-presentation-in-movies-has-the-of-bechdel-test-passing-movies-a34290f26412

Outspoken, The. (June 10th, 2015). Women in horror – are horror movies inherently misogynistic, or are they just misunderstood? https://medium.com/@Ghill_deRozario/are-horror-movies-inherently-misogynistic-or-are-they-just-misunderstood-76990d8753aa

Sieczkowski, Cavan. (June 3rd, 2016). Why the internet is disturbed by this "X-Men" poster. https://www.huffpost.com/entry/jennifer-lawrence-x-men-poster-strangle_n_57507594e4b0c3752dccdcbd

Other Resources That Influenced This Writing

Some of these were not directly referenced, and others were used in multiple chapters and therefore could not fit neatly into one of the above headings. This section also includes books, podcasts, and YouTube channels that I found helpful.

Almwaka, Majdoulin. (2021). Complex female agency and the "final girl" trope, and the subversion and reaffirmation of patriarchy: the cases of western & MENA horror films. *Journal of International Women's Studies*: Vol. 24: Iss. 3, Article 5. https://vc.bridgew.edu/jiws/vol24/iss3/5

Berlatsky, Noah. (July 29th, 2016). *Why violence against women in film is not the same as violence against men.* The Guardian. https://www.theguardian.com/film/2016/jul/29/the-killing-joke-batgirl-violence-against-women-men

Bradley, S. A. (2018). *Screaming for pleasure: How horror makes you happy and healthy.* Coal Cracker Press.

Cheung, Kylie. (October 10th, 2022). *Sexist disbelief is taking over the horror genre.* Jezebel. https://jezebel.com/sexist-disbelief-is-taking-over-the-horror-genre-1849608771

Clover, C. J. (1992). *Men, women, and chainsaws: Gender in the modern horror film.* (With a new preface by the author). Princeton University Press.

Crump, Andy. (May 11th, 2019). How *'The Witch'* accidentally launched a horror movement. The Hollywood Reporter. https://www.hollywoodreporter.com/movies/movie-features/how-witch-accidentally-sparked-elevated-horror-trend-1208008/

Dead Meat. (2017-Present). Youtube. https://www.youtube.com/@DeadMeat/videos

Dray, Kayleigh. (Accessed 12/18/2022). *Scream 5*: how the *Scream* franchise has been killing off sexist horror tropes for 25 years. Stylist. https://www.stylist.co.uk/life/scream-horror-movies-sexism-feminism-neve-campbell-courteney-cox/412951

Gilbey, Ryan. (October 15th, 2021). Jamie Lee Curtis: 'My biggest roles were to do with my body, my physicality, my sexuality'. https://theguardian.com/film/2021/oct/15/jaime-lee-curtis-my-biggest-roles-were-to-do-with-my-body-my-physicality-my-sexuality

Grant, B. K. (Ed). (1996). *The dread of difference: Gender and the horror film* (2nd ed.) University of Texas Press. doi: 10.7560/771376

Hafdahl, M. & Florence, K. (2020). *The science of women in horror: The special effects, stunts, and true stories behind your favorite fright films.* Skyhorse Publishing.

Hankins, Sarah. (2019). "Torture the women": a gaze at the misogynistic machinery of scary cinema". *Copley Library Undergraduate Research Awards.* https://digital.sandiego.edu/library-research-award/1

Jacobs, Matthew. (October 31st, 2018). The horror returns: inside the shared legacies of 'Halloween' and 'Suspiria'. https://www.huffpost.com/entry/halloween-suspiria-shared-legacies_n_5bcb794ce4b0a8f17eed0015

Janisse, J. & Rebecca, C. (Hosts). (2018-Present). *Dead Meat.* Independently produced.

Kennedy, Caitlin. (January 28th, 2020). The final girl, redefined: the role of women in horror is evolving. https://www.themarysue.com/final-girl-role-of-women-in-horror-evolving/

King-Miller, Lindsay. (September 17th, 2019). *A love letter to the girls who die first in horror films.* Electric Literature. https://electricliterature.com/a-love-letter-to-the-girls-who-die-first-in-horror-films/

Kisner, Logan Ashley. (December 15th, 2020). *What makes a good feminist horror film?* An Injustice Mag. https://aninjusticemag.com/what-makes-a-good-feminist-horror-film-8d53f228b563

McKendry, R. & Kane, E. (Hosts). (2019-Present). *Colors of the Dark.* Produced by Fangoria.

Monroe, Rachel. (2020). *Savage appetites: True stories of women, crime, and obsession.* Simon & Schuster.

Park, Michelle. "The Aesthetics and Psychology Behind Horror Films" (2018). *Undergraduate Honors College Theses 2016.* 31. https://digitalcommons.liu.edu/post_honors_theses/31

Romano, Aja. (October 26th, 2021). *Scream* broke all the rules of horror – then rewrote them forever. Vox. https://www.vox.com/22634481/scream-influence-horror-genre-wes-craven-new-nightmare-get-out

Skal, D. J. (1993). *The monster show: A cultural history of horror* (Revised edition). Farrar, Straus and Giroux.

Stipidis, Julieann. (March 5th, 2020). She's not imagining things: 'The Invisible Man' and horror's history of believing women. https://bloody-disgusting.com/editorials/3607888/shes-not-imagining-things-invisible-man-horrors-history-believing-women/

Subissati, A. & West, A. (Hosts). (2012-Present). *The Faculty of Horror.* Independently produced.

The Take. (2021, July 27). *Jennifer's Body and the horrific female gaze.* [Video]. Youtube. https://www.youtube.com/watch?v=X7Twg8rG2HI

The Take. (2021, June 12). *The female gaze – yes it can exist.* [Video]. Youtube. https://www.youtube.com/watch?v=eCPD7Mi9504

The women missing from the silver screen and the technology used to find them. Google. https://about.google/main/gender-equality-films/

VB, Lizzy. (February 1st, 2020). *Why horror is the most feminist genre.* Morbidly Beautiful. https://morbidlybeautiful.com/horror-feminist-genre/

Yang, Haiyang and Zhang, Kuangjie. (October 26th, 2021). The psychology behind why we love (or hate) horror. https://hbr.org/2021/10/the-psychology-behind-why-we-love-or-hate-horror

Younger, Beth. (June 25th, 2017). Women in horror: victims no more. https://theconversation.com/women-in-horror-victims-no-more-78711

Movies Referenced

28 Days Later (2002)	Danny Boyle
Abbott and Costello Meet Frankenstein (1948)	Charles Barton
Accepted (2006)	Steve Pink
Alien (1979)	Ridley Scott
Aliens (1986)	James Cameron
Alien³ (1992)	David Fincher
Alien: Resurrection (1997)	Jean-Pierre Jeunet
American Psycho (2000)	Mary Harron
A Fool and His Money (1912)	Alice Guy-Blaché
A Nightmare on Elm Street (1984)	Wes Craven
Arachnophobia (1990)	Frank Marshall
As Above, So Below (2014)	John Erick Dowdle
Attack of the 50 Foot Woman (1958)	Nathan Juran
Attack of the Crab Monsters (1957)	Roger Corman
The Autopsy of Jane Doe (2016)	André Øvredal
The Babadook (2014)	Jennifer Kent
Behind the Mask: The Rise of Leslie Vernon (2006)	
	Scott Glosserman
Black Christmas (1974)	Bob Clark
Black Sunday (1960)	Mario Bava
The Blair Witch Project (1999)	Eduardo Sánchez
	Daniel Myrick
The Blob (1958)	Irwin Yeaworth
	Russell Doughten

The Birds (1963)	Alfred Hitchcock
The Brave One (2007)	Neil Jordan
Bride of Frankenstein (1935)	James Whale
The Cabbage Fairy (1896)	Alice Guy-Blaché
Cabin Fever (2002)	Eli Roth
The Cabin in the Woods (2012)	Drew Goddard
The Cabinet of Dr. Caligari (1920)	Robert Wiene
Cam (2018)	Daniel Goldhaber
Candyman (1992)	Bernard Rose
Cannibal Holocaust (1980)	Ruggero Deodato
Captain Marvel (2019)	Anna Boden
	Ryan Fleck
Carrie (1976)	Brian De Palma
Carrie (2013)	Kimberly Pierce
The Cat and the Canary (1927)	Paul Leni
Cloverfield (2008)	Matt Reeves
The Conjuring (2013)	James Wan
The Conjuring 2 (2016)	James Wan
The Conjuring: The Devil Made Me Do It (2021)	
	Michael Chaves
The Craft (1996)	Andrew Fleming
Critters (1986)	Stephen Herek
Dark Water (2005)	Walter Salles
Deadpool (2016)	Tim Miller
Deliverance (1972)	John Boorman
The Descent (2007)	Neil Marshall
Doctor Sleep (2019)	Mike Flanagan
Everest (2015)	Baltasar Kormákur

The Exorcist (1973)	Walter Friedkin
The Exorcism of Emily Rose (2005)	Scott Derrickson
The Eye (2008)	David Moreau
	Xavier Palud
Final Destination (2000)	James Wong
Forgetting Sarah Marshall (2008)	Nicholas Stoller
Frankenstein (1931)	James Whale
Freaks (1932)	Tod Browning
Friday the 13th (1980)	Sean S. Cunningham
Frozen (2010)	Adam Green
Funny Games (2007)	Michael Haneke
Gaslight (1944)	George Cukor
Gerald's Game (2017)	Mike Flanagan
Get Out (2017)	Jordan Peele
The Ghost and Mrs. Muir (1947)	Joseph L. Mankiewicz
Ghostbusters (2016)	Paul Feig
Ginger Snaps (2000)	John Fawcett
The Green Inferno (2013)	Eli Roth
Gremlins (1984)	Joe Dante
The Grudge (2004)	Takashi Shimizu
Halloween (1978)	John Carpenter
Hellraiser (1987)	Clive Barker
Hellraiser (2022)	David Bruckner
Hereditary (2018)	Ari Aster
The Hills Have Eyes (1977)	Wes Craven
The Hitchhiker (1953)	Ida Lupino
Hollow Man (2000)	Paul Verhoeven
Hostel (2005)	Eli Roth

Hostel: Part II (2007)	Eli Roth
Hostel: Part III (2011)	Scott Speigel
House of Wax (2005)	Jaume Collet-Serra
The Hunchback of Notre Dame (1923)	Wallace Worsley
Hush (2016)	Mike Flanagan
I Spit on Your Grave (1978)	Meir Zarchi
I Spit on Your Grave (2010)	Steven R. Monroe
I Spit on Your Grave: Déjà Vu (2020)	Meir Zarchi
Incident in a Ghostland (2018)	Pascal Laugier
Insidious (2010)	James Wan
Invasion of the Body Snatchers (1956)	Don Siegel
Invisible Man (2020)	Leigh Whannel
It Follows (2014)	David Robert Mitchell
Jaws (1975)	Steven Spielberg
Jaws 2 (1978)	Jeannot Szwarc
Jaws 3-D (1983)	Joe Alves
Jaws: The Revenge (1987)	Joseph Sargent
Jennifer's Body (2009)	Karyn Kusama
Jurassic Park (1993)	Steven Spielberg
The Last House on the Left (1972)	Wes Craven
The Last House on the Left (2009)	Dennis Ilidas
The Last Performance (1929)	Paul Fejos
Le Caverne Maudite (1898)	George Méliès
Le Manior du Diable (1896)	George Méliès
Le Papillon Fantastique (1909)	George Méliès
Les Vampires (1912)	Louis Feuillade
The Lighthouse (2019)	Robert Eggers
Lipstick (1976)	Lamont Johnson

The Lord of the Rings Trilogy (2001-2003)

Peter Jackson

Malignant (2021)   James Wan

The Man Who Laughs (1928)   Paul Leni

Martyrs (2008)   Pascal Laugier

M3gan (2022)   Gerard Johnstone

MFA (2017)   Natalia Leite

Midsommar (2019)   Ari Aster

Misery (1990)   Rob Reiner

The Monkey's Paw (1948)   Norman Lee

Ms. 45 (1981)   Abel Ferrara

The Mummy (1932)   Karl Freund

The Mummy (1999)   Stephen Sommers

The Mummy (2017)   Alex Kurtzman

Night of the Living Dead (1968)   George A. Romero

Nosferatu (1922)   F. W. Murnau

Ocean's 8 (2018)   Gary Ross

Oculus (2013)   Mike Flanagan

The Omen (1976)   Richard Donner

One Missed Call (2008)   Éric Valette

The Phantom of the Opera (1925)   Lon Chaney, Rupert Julian,

Edward Sedgwick,

Ernst Laemmie

The Pit and the Pendulum (1913)   Alice Guy-Blaché

Plan 9 From Outer Space (1959)   Ed Wood

Prometheus (2012)   Ridley Scott

Promising Young Woman (2020)   Emerald Fennell

Psycho (1960)   Alfred Hitchcock

Pulse (2006)	Jim Sonzero
The Purge (2013)	James DeMonaco
Ready or Not (2019)	Matt Bettinello-Olpin
	Tyler Gillett
Revenge (2017)	Coralie Fargeat
The Ring (2002)	Gore Verbinski
Rosemary's Baby (1968)	Roman Polanski
Saw (2004)	James Wan
Scary Movie (2000)	Keenan Ivory Wayans
Scary Movie 2 (2001)	Keenan Ivory Wayans
Scary Movie 3 (2003)	David Zucker
Scary Movie 4 (2006)	David Zucker
Scary Movie 5 (2013)	Malcolm D. Lee
Scream (1996)	Wes Craven
Shaun of the Dead (2004)	Edgar Wright
The Shining (1980)	Stanley Kubrick
Shutter (2008)	Masayuki Ochiai
Signs (2002)	M. Night Shyamalan
The Silence of the Lambs (1991)	Jonathan Demme
Silent Night, Bloody Night (1972)	Theodore Gershuny
The Sixth Sense (1999)	M. Night Shyamalan
The Strangers (2008)	Bryan Bertino
The Strangers: Prey at Night (2018)	Johannes Roberts
Teeth (2007)	Mitchell Lichtenstein
The Texas Chainsaw Massacre (1974)	Tobe Hooper
The Texas Chainsaw Massacre (2006)	Jonathan Liebesman
The Thing (1982)	John Carpenter
Thirteen Women (1932)	George Archainbaud

The Tingler (1959)	William Castle
Tremors (1990)	Ron Underwood
Umma (2022)	Iris Shim
The Uninvited (2009)	Thomas Guard
	Charles Guard
Urban Legend (1998)	Jaime Blanks
Us (2019)	Jordan Peele
When A Stranger Calls (1979)	Fred Walton
White Zombie (1932)	Victor Halperin
The Witch (2015)	Robert Eggers
Wolf Creek (2005)	Greg McLean
Wonder Woman (2017)	Patty Jenkins
X (2022)	Ti West
X-Men Apocalypse (2016)	Bryan Singer
You're Next (2011)	Adam Wingard
Zombieland (2009)	Ruben Fleischer

Made in the USA
Middletown, DE
08 June 2024

55293255R00145